MANAGEMENT OF THE ENVIRONMENTAL IMPACT OF AIRPORT OPERATIONS

MANAGEMENT OF THE ENVIRONMENTAL IMPACT OF AIRPORT OPERATIONS

H.G. VISSER,
S.J. HEBLY
AND
R.A.A. WIJNEN

Nova Science Publishers, Inc.
New York

For permission to use material from this book please contact us:
Telephone 631-231-7269; Fax 631-231-8175
Web Site: http://www.novapublishers.com

LIBRARY OF CONGRESS CATALOGING-IN-PUBLICATION DATA

Visser, H. G.

Management of the environmental impact at airport operations / H.G. Visser, S.J. Hebly, and R.A.A. Wijnen.

p. cm.

ISBN 978-1-60456-941-4 (softcover)

1. Airport noise. 2. Noise control. 3. Airports--Environmental aspects. I. Hebly, S. J. II. Wijnen, R. A. A. III. Title.

TL725.3.N6V57 2008

387.7'4042--dc22 2008026002

Published by Nova Science Publishers, Inc. ✢ New York

CONTENTS

Preface		**ix**
Introduction		**ix**
Chapter 1	Environmental Impacts Associated with Airport Operations	**1**
Chapter 2	Integrated Environmental Management Concept	**15**
Chapter 3	Optimization of Noise Abatement Trajectories	**21**
Chapter 4	Environmental Planning at the Strategic Level	**57**
Chapter 5	Environmental Planning at the Tactical/Operational Level	**67**
Conclusion		**75**
Acknowledgments		**77**
References		**79**
Index		**83**

PREFACE

Today's airports are at times unable to handle the air traffic demand. The busiest airports are already saturated, and there are political and environmental difficulties associated with any further airport expansion. In view of the anticipated growth in air traffic demand, there is a clear need for economically beneficial capacity improvements in an environmentally responsible manner. However, the required capacity growth cannot be achieved by relying on existing technologies, policies and procedures. To provide solutions for environmentally induced capacity bottlenecks, our research is aimed at the development of a new integrated concept for managing the environmental impact of flight operations into and out of airports. The set of fully integrated noise management tools that we envision includes interrelated tools at the strategic level (annual/seasonal noise allocation planning), the tactical/operational level (sequencing and scheduling of flights and separation assurance) and the trajectory level (selection of noise-optimized routes and flight profiles). The proposed integrated environmental management tool provides decision support to air traffic controllers to enable traffic management on the basis of throughput efficiency and safety in concert with noise exposure and emission considerations. The objective of this chapter is to outline the envisaged integrated environmental management concept, and to summarize our research efforts related to the main enabling capabilities (tools) that underlie this concept.

INTRODUCTION

Due to the continuing growth in air traffic demand and the rising level of urbanization around many airports, airports across the globe are increasingly confronted with the need to deal with the impact of noise and emissions on the quality of life in the surrounding residential communities. Many airports already face severe difficulties in balancing airport expansion requirements to meet the growing traffic demand, and societal demands for a reduced environmental impact.

To mitigate the impact of aircraft noise and other environmental impacts, an array of measures has been developed and implemented at airports located close to sensitive communities. Although societal concerns about third party risk and air quality issues are on the rise, noise impact remains the most significant environmental concern. The possibilities to mitigate the effect of airport noise include:

- Land-use and noise compatibility policies, including zoning restrictions, sound insulation programs and land and property acquisition
- Airport access restrictions, including operating quota, noise budgets, (nighttime) curfews and restrictions on the operation of certain (noisy) types of aircraft
- Noise monitoring and reporting
- Airport and airspace design features, including high speed runway exits, sound barriers, improved lay-out of the taxiway system, displaced runway threshold and raised glide slope angle
- Preferential or rotational use of runway configurations
- Noise abatement procedures

- Restrictions on surface operations, including engine run-up and APU operating restrictions, restrictions on the use of reverse thrust and towing of aircraft
- Community relations and public involvement programs
- Noise-related landing fees or penalties

In the early days of noise management, airports mostly adopted measures that are reactive in nature (e.g., noise monitoring and reporting), but over time airports started to take a more proactive and strategic approach to the management of aircraft noise. Proactive, strategic measures, such as land zoning restrictions, are preventive in nature and have a relatively high potential to produce noise benefits. It needs to be noted that some measures are more widespread than others. For example, at 87% of the major European airports, noise abatement procedures are in use, whilst at only 5% of the observed European airports, noise budget restrictions are in place [Anton Cruz, 2000].

One of the most noise-sensitive major airports in Europe is Schiphol airport, located near Amsterdam in The Netherlands. In the last few decades, proposals for new infrastructure developments or expanded operations at Schiphol airport have consistently met with significant community opposition. Spurred by a vocal public debate, the noise nuisance problem at Schiphol airport has been elevated to a major political issue in The Netherlands. To control the noise resulting from flight operations and to mitigate its worst effects, Schiphol airport and (national and local) authorities have adopted and legislated an extensive set of measures, comprising virtually all of the above items. This has resulted in a situation where noise constraints, and not so much the physical capacity (e.g., runway capacity), determine Schiphol's ability to service (future) traffic demand. This situation is certainly not unique to Schiphol; environmental issues, particularly the impact of aircraft noise, constitute the most significant constraint on the growth potential of airport businesses for most large airports in the developed world [ACI, 2001].

One of the most widely employed options to reduce the noise impact is to re-shape arrival and departure trajectories into and out of an airport. For this purpose, noise-abatement routes and procedures have been designed and implemented on a fairly wide scale. At many airports around the world, including Amsterdam airport, ICAO Noise Abatement Take-off procedures (occasionally modified to match local conditions) have become commonplace [Clarke, 2000]. With respect to the approach phase, a great variety of procedures exists that might be classified as noise abatement procedures. At Amsterdam airport, the so-called Continuous Descent Approach (CDA) is currently in use during the noise sensitive night-time hours. The applied procedure, enabled by modern guidance and navigation technology such as area navigation (RNAV) and flight management system (FMS), allows aircraft to descend continuously from high altitude without any level flight segment at low altitude. The higher flight path combined with lower engine thrust helps to reduce noise exposure in the communities surrounding the airport. Unfortunately, however, with the present-day state of technology, the predictability of arrival time over the runway threshold under such procedures is rather poor. As a consequence, fairly large separation intervals in the arrival sequence are currently required, which obviously limits the application of the CDA noise abatement procedure to low-demand hours only.

To achieve community-wide noise reduction objectives, a minimization of the noise impact of each individual flight movement considered in isolation is not sufficient. The reason for this is that noise exposure is “additive” in nature. Indeed, the true community noise impact is based on the aggregation and distribution of all flight movements for a given time period (e.g., a year, or a season). In many countries noise regulations provide standards for the acceptable level of annual noise exposure for persons living in the vicinity of airports. In the Netherlands, for example, the noise regulations currently applicable to Schiphol airport stipulate limits for the total annual noise volume and for the average annual noise exposure at a fixed number of predefined locations in the vicinity of the airport. The "environmental capacity" of Schiphol is reached when the adverse environmental consequences of its operations approach these regulatory limits. It is evident that efficient community noise management calls for an optimal allocation and distribution of the annual flight movements over runways and routes.

Although there is a close relationship between Air Traffic Control (ATC) operations and environmental impact, air traffic controllers traditionally manage traffic on the basis of efficiency and safety, without explicitly taking into account environmental concerns. Decision support tools for arrival and departure management that are currently in use or in development are typically designed to offer assistance to terminal area controllers in establishing safe and efficient sequences, with the sole aim to maximize the use of available runway capacity. To date, no real-time decision support is in place that helps controllers in determining route, schedule and sequence changes on a per aircraft basis to affect long-term, cumulative noise exposure.

Whilst many of the traditional measures have had a marked effect on the level of noise impact on surrounding communities, the rate of growth in aircraft operations is outstripping the rate of operational improvements with the result that noise continues to remain a major source of annoyance. In addition, public concerns about local air quality issues are rapidly growing, bringing about the need for introducing measures to mitigate air pollution. To address these problems, research is currently focusing on identifying and exploring new options to resolve environmentally induced capacity bottlenecks. One particular initiative to increase environmental capacity that is currently being developed can be loosely qualified as a "collaborative" approach. It follows two main strands.

The first strand is a collaborative approach between airport and community, with the aim to improve environmental capacity by building community tolerance to noise. This is to be achieved by identifying aspects of airport operations that can be managed in concert with the affected communities. This particular approach is based on the recognition that individuals that live in the residential areas close to an airport, are to a certain extent stressed by the fact that they are exposed to events that personally affect them, but over which they have (or perceive to have) little to no control. Through information, consultation and empowerment of the affected communities, environmentally induced anxiety symptoms are likely to be attenuated, with consequent benefits for environmental capacity [Maris, 2007].

The second strand of research seeks to increase the involvement of air traffic controllers in the noise mitigation process. One way to achieve this is through the development of decision support tools that factor aircraft noise in the terminal area traffic management process to achieve long-term community noise mitigation goals. While the targeted community tolerance building initiatives help to increase environmental capacity, the envisioned controller support tools essentially provide the means to make best use of the available environmental capacity. Although we believe that a balanced blend of both types of initiatives is likely to lead to the most effective approach to noise management, our research efforts are primarily aimed at the development of decision support tools that integrate noise mitigation and air traffic control decision making. However, some of fundamental ideas that underlie most new community tolerance building initiatives, such as the use of alternative noise descriptors and metrics to make the employed noise management principles easier to understand, have also been embraced in our work.

The objective of this Chapter is to outline our proposal for a new integrated environmental management concept, and to summarize our research efforts related to the main enabling capabilities (tools) that underlie this concept. The set of fully integrated environmental management tools that we envision includes interrelated tools at the strategic level (annual/seasonal noise allocation planning), at the tactical-/operational level (sequencing and scheduling of flights and separation assurance) and at the trajectory level (noise-optimized routes and flight paths). Although the numerical examples that we present pertain to a particular airport, viz. Schiphol airport, our work actually focuses on the development of a generic methodology that can be applied to any given airport.

Section II reviews some basic concepts used in the assessment of environmental impacts due to airport operations. In Section III, we describe the proposed integrated environment management concept and we provide an overview of the main building blocks of its architecture. In Section IV we discuss environmental impact assessment and mitigation at the trajectory level and describe NOISHHHH, a tool for the design and synthesis of advanced noise abatement procedures for arrivals and departures. In section V, we focus on environmental planning at the strategic level; we outline a tool called SNAP that optimally allocates and distributes the annual flight demand over runways and routes, according to a set of user-specified criteria. Finally, in Section VI we report on the work in progress towards the development of a decision support tool at the tactical-/operational level and we discuss the interrelationships and interactions among the tools at the various levels of aggregation

Chapter 1

Environmental Impacts Associated with Airport Operations

Noise Descriptors

To describe aircraft noise, a large variety of acoustic descriptors is in use [Ruijgrok, 2004]. It is of paramount importance that when an aircraft noise descriptor is selected, it matches the needs of the issue being examined. To make such judgments requires a good understanding of how noise is quantified and measured.

In the literature, noise and sound are often used interchangeably, although in principle noise merely represents "unwanted" sounds. Also in this Chapter we will use the two terms interchangeably. From a physical point of view, sound is a phenomenon consisting of minute vibrations, generated by a source – an aircraft, in our case – and which travel through the atmosphere to reach our ears, where they create pressure fluctuations that can be sensed. The perception and measurement of sound involves two basic physical characteristics: (i) intensity (volume), which is a measure of the acoustic energy, usually expressed in terms of a logarithmic unit of measure called "Decibel" (dB), and (ii) frequency (pitch) of the sound vibrations, typically expressed in Herz (Hz).

In the case of airport noise, a so-called A-weighted adjustment is generally applied to sound level measurements. A-weighting corrects sound levels for the changing sensitivity of the human ear with frequency. To this end, frequencies outside the high-sensitivity range (2,000-6,000 Hz) are de-emphasized. Other forms of weighting are also in use; for example, so-called tone-corrected perceived noise levels are used in connection with aircraft certification for noise.

The acoustic descriptors that will be referred to here are all based on A-weighted sound levels and are expressed in terms of the unit of measure dB(A).

The most commonly used A-weighted noise descriptors are in practice subdivided into: (i) single-event noise descriptors, associated with a single flight movement, and (ii) cumulative noise descriptors, which seek to capture the aggregate effect of all movements over a specified period.

Within the family of single-event noise descriptors, three types of descriptors can be distinguished: (i) exposure-based descriptors, that represent the sound exposure for a finite time period, (ii) maximum noise level descriptors, that represent the maximum value in the sound pressure level history of a single flight, and (iii) time-above descriptors, that represent the duration that a time-varying sound level is above a given threshold. In the US, in Europe and in much of the rest of the world, sound exposure level (SEL) and maximum sound level ($L_{A\ max}$) are the predominant exposure-based and maximum-level descriptors, respectively.

Cumulative noise descriptors are used to quantify at a given location the combined impact of the loudness of the individual noise events and of the frequency of occurrence of these events. Due to the fact that it reliably correlates with health and welfare effects, the A-weighted equivalent sound level (L_{eq}) is considered one of the most appropriate community noise descriptors. The equivalent sound level L_{eq} represents the time average of the total sound energy over a specified period. More specifically, when during a time period T a total of N flights take place, the equivalent sound level L_{eq} can be expressed as:

$$L_{eq} = 10 \log \left[\frac{1}{T} \sum_{i=1}^{N} 10^{SEL_i/10} \right] ,$$

where SEL_i is the sound exposure level of the ith flight movement. The L_{eq} descriptor is generic in nature and several special variants have been developed that incorporate weightings of the single event levels depending on the time of day or night at which they occur. Three of the special variants are particularly important: (i) day-night average sound level L_{dn} (DNL), which features a adjustment for the nighttime events (10-dB(A) weighting), (ii) day-evening-night average sound level L_{den}, which features adjustments for the evening events (5-dB(A) weighting) and for nighttime events (10-dB(A) weighting), and (iii) L_{night}, which is the long term average sound level determined over all the night periods of a year. Whilst L_{dn} is employed as a standard metric in the US, L_{den} and L_{night} are the current European noise level indicators to describe environmental noise during the day and night, respectively.

The acoustic community noise descriptors referred to in this sub-section generally have in common that they do not provide noise information in a way that an average member of the public can relate to or can readily understand. To facilitate an informed dialogue regarding environmental issues between airports and their neighboring communities, there is clear need for descriptors that are less technical and more transparent to non-experts. In response to these needs several attempts to devise new descriptors are being undertaken [DOTARS, 2003]. A good example of such an innovative descriptor is the N70 index, also called the NA70 index (Noise Above 70 db(A)) [DOTARS, 2003]. The noise contours on an N70 chart indicate the number of aircraft noise events louder than 70 dB(A) which occurred on the average day during the period covered by the chart. In contrast to the acoustic community noise descriptors, which provide a 24-hour logarithmic average of noise levels, the N70 index aggregates the single-event noise data in a form that is comprehensible to the non-expert. The N70 descriptor provides context to both the peak noise and how often loud noise events are present within the area of study. Although the N70 contours may be easier to understand, they still don't allow to fully address community concerns. For example, the N70 descriptor fails to convey the durational aspect of a noise event. This limitation can be partlially overcome by complementing the N70 descriptor by a temporal descriptor such as Time Above 70 dB(A) (TA70).

Noise Effects

For the purpose of land-use planning and noise mitigation programs, it is useful to have a set of relationships that show which annoyance level is associated with a given noise exposure level in a community. In essence such so-called dose-response relationships combine the physical measurement of sound with a scientific assessment of the community perception of sound. To establish dose-response relationships, a variety of studies has been conducted across the globe.

One of the most widely used sets of dose-response relationships (for various modes of transport) is the well-known set of curves from the TNO data collection [Miedema, 1998]. This data set comprises annoyance and disturbance data compiled from about 50 different studies on aircraft, road and rail noise that were conducted in Australia, Europe and North America. In the results, the percentage of the people that is annoyed or highly annoyed by the noise exposure is recorded against the noise exposure level. The noise exposure descriptors that are used in these particular dose-response relationships are L_{den} and L_{dn}, due to the finding that they best predict annoyance from aircraft noise [Miedema, 2000]. One set of curves derived in this study is presented in Figure 1. The obtained curves, for exposure between 45 and 75 dB(A) in both metrics, can be approximated accurately with third order polynomials. For aircraft noise, the polynomials approximations for percentage of persons annoyed ($\%A_{dn}$ and $\%A_{den}$) are given, respectively, by:

$$\%A_{dn} = 1.460e^{-5}\,(L_{dn} - 37)^3 + 1.511e^{-2}\,(L_{dn} - 37)^2 + 1.346(L_{dn} - 37)$$

$$\%A_{den} = 8.588e^{-6}\,(L_{den} - 37)^3 + 1.777e^{-2}\,(L_{den} - 37)^2 + 1.221(L_{den} - 37)$$

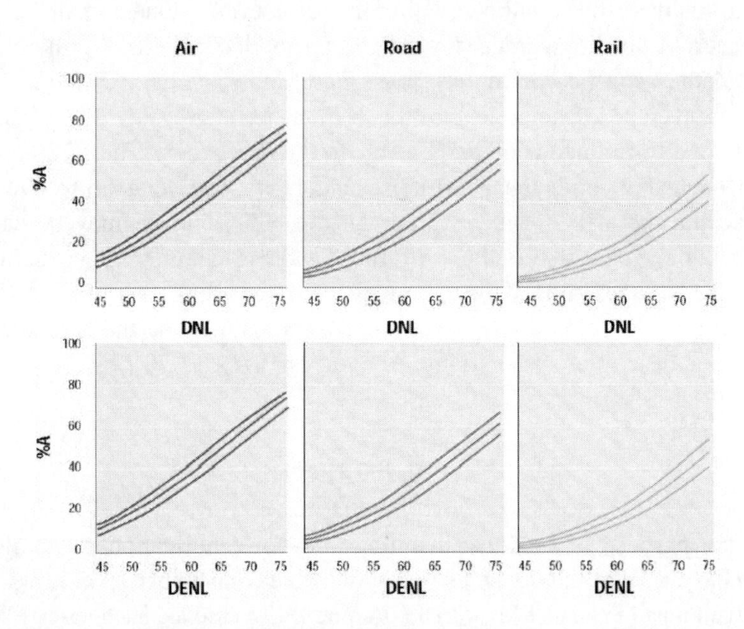

Figure 1. Annoyance due to air, road and rail noise in Ldn (LDN) as well as in Lden (LDEN). All sub-figures show the average annoyance and the 95% confidence interval. Image taken from [Miedema, 1998].

In addition to the percentage of persons annoyed, also the percentage of persons highly annoyed is recommended in [EC, 2002] as a descriptor of noise annoyance in a population. Although the numerical values obtained for the annoyance descriptors cannot be used as if they would represent actual numbers of people that are annoyed by aircraft noise, they are very useful in strategic assessments, where the emphasis is not so much on absolute performance, but rather on comparing the relative performance of various alternatives [Miedema, 1998]. They can be used, for example, in cost-benefit evaluations or as a performance objective in optimization studies.

In the US, the so-called “updated Schultz curve” remains the primary source of empirical dosage-effect information to predict community response to aircraft noise [FICAN, 1997]. A comparison of the Schultz and Miedema annoyance curves is shown in Figure 2.

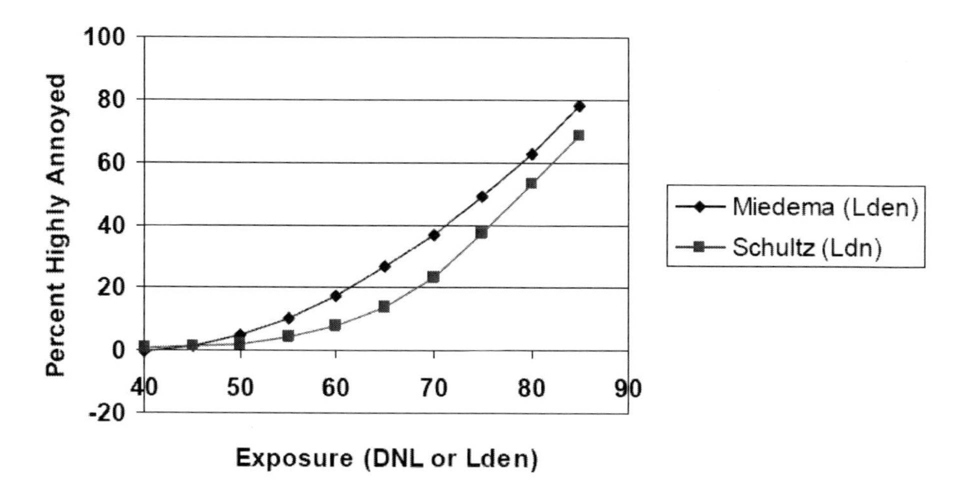

Figure 2. Comparison of Relationships between Noise Level and Percent of People ‘Highly Annoyed; Image taken from [Eagan, 2006].

It has been long recognized that the human tolerance to noise nuisance is much higher during the day than during the night. In other words, sounds that people would ignore during the day become annoying at night [Ruijgrok, 2004]. In addition to social surveys, physiological monitoring of sleepers under both field and laboratory conditions has been carried out to study the effects of noise on sleep. Many of these studies on sleep interference have varying conclusions as to the effect of noise on sleep. One of the most widely adopted sleep disturbance dose-response relationships has been proposed by the Federal Interagency Committee on Aircraft Noise [FICAN, 1997], a standard-setting organization in the US. The "FICAN 1997" curve, shown in Figure 3, is based on the combined data and conclusions of three different field studies and provides a conservative dose-response relationship. Indeed, this relationship represents a worst-case bound on the number of people likely to awake. The FICAN 1997 curve can be represented by the following relationship for the percentage of the exposed population expected to be awakened (%Awakenings) as a function of the exposure to single event noise levels expressed in terms of sound exposure level:

$$\%Awakenings = .0087\ (SEL_{indoor} - 30\ )^{1.79}\ ,$$

where SEL_{indoor} is defined to be the Sound Exposure Level (dB(A)) which occurs indoors. FICAN cautions that the dose-response relationship presented here solely relies on behavioral awakening as the indicator of sleep disturbance.

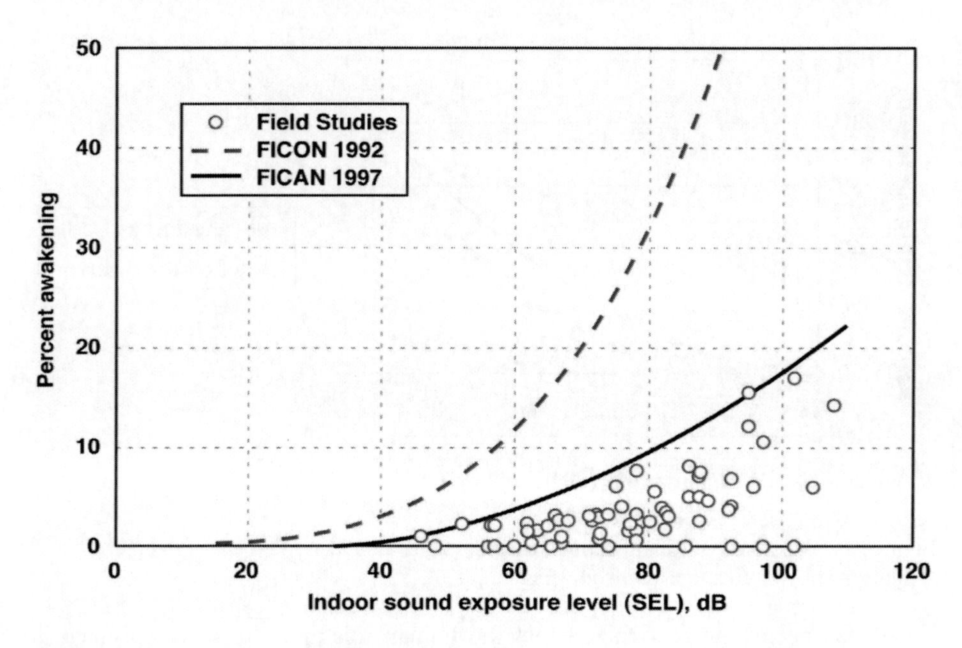

Figure 3. FICAN proposed sleep disturbance dose-response relationship; Source: [FICAN, 1997].

One significant weakness of the FICAN 1997 relationship is that it predicts expected percent awakening for a single event only; it does not predict what happens over the course of an entire night, when a community may be exposed to multiple noise events. In [Eagan, 2006] the FICAN 1997 relationship is extended through the use of probability calculations to predict percent of population awakening from repeated aircraft noise events. Note that this approach assumes that all noise events are independent. This may not always be a reasonable assumption, and the predicted chance of awakening is therefore likely to be conservative.

It is important to realize that noise annoyance is not only determined by acoustic factors, there also many social, economic, cultural and attitudinal factors that affect peoples perception of what is an acceptable level of disturbance (e.g. tolerance is lower amongst more affluent communities). Unfortunately, at the present time, there are no quantitative methods in use that relate the level and acceptability of the perceived disturbance to socio-economic, cultural and other non-acoustic factors. It is for this reason that acoustic descriptors of noise exposure remain as the primary regulatory indicators for noise disturbance.

NOISE MODELING

In order to estimate single-event and cumulative metrics of noise exposure at observer locations around the airport and to estimate the associated contours, a large number of computer-based noise prediction models has been developed. The selection of the noise model that is most appropriate in a particular study largely depends on the reasons for performing the noise analysis.

The Aircraft Noise Prediction Program (ANOPP), for example, is a high-fidelity noise analysis program that allows for the detailed computation of noise as generated by a single aircraft, taking into account the contributions of the most significant individual components of the noise source [Hough, 1996]. The ANOPP model is most appropriate to assess and display the noise impact of a specific trajectory that is flown. A related development concerns NOISIM, a tool that combines a noise model (based on ANOPP) and a real-time aircraft simulator, enabling to investigate trade-offs between noise impact and aircraft performance [Clarke, 2000]. NOISIM implicitly takes into account the effect of wind or other atmospheric conditions on aircraft performance and noise propagation. To assess noise impacts associated with aircraft routings over large regions and airports, the Noise Integrated Routing System (NIRS) has been developed [Metron Aviation, 2007]. The NIRS tool, which is currently the United States' Federal Aviation Administration's (FAA) standard regional noise model, is a program that efficiently addresses the large scale requirements of modeling aircraft noise effects over large regions of airspace.

One of the most widely used tools for assessing the changes in noise impact resulting from revised routings or alternative flight paths is the Integrated Noise Model (INM) [FAA OEE, 1997]. INM, which has been developed in several successive versions, has been the FAA's official methodology for noise impact assessment in the vicinity of civilian airports since 1978. Over time INM has been adopted as the standard tool for noise impact assessment by many countries across the globe, to the point that it has become a virtual international standard.

To model the movement of an aircraft in three-dimensional (3D) space, INM describes a flight path as a sequence of straight-line segments of finite length. To compute the sound exposure level at a given observer point on the ground, the contribution of each segment of the flight-path to the overall result has to be taken into account. The INM procedure for determining the sound exposure level, at any specific observer location, is to select appropriate sound levels from a Noise-Thrust-Distance (NTD) table corresponding to the distances from aircraft to observer. The NTD data contained in the INM database represent the noise exposure levels for specific reference conditions for each aircraft type. To allow for differences between the actual conditions and the reference conditions specified for the NTD tables, a number of noise level adjustments need to be made.

The noise at the observer points from a single aircraft flyover depends on a number of factors. Principal among these are the types of aircraft and engines; airspeed and power setting; aircraft configuration (flaps, undercarriage); the distances from the points concerned to the various flight path segments; and local topography and weather, affecting sound propagation.

To assess community noise impact a significant amount of site-specific data on operations at the considered airport or region is required, including runway configurations and frequency of use, assignment of aircraft types to arrival and departure runways, assignment of aircraft to arrival and departure flight paths, and geometric characteristics of the runways.

In comparison to models such as ANOPP, the INM omits a great deal of detail. Moreover, INM relies on a variety of simplifying assumptions, rendering its estimates of noise exposure far from accurate. Nevertheless, the INM has established itself worldwide as the foremost strategic tool for long-term, average-value community noise estimation, primarily by providing a common ground among airports for assessing the relative noise performance.

In order to be able to produce noise information that is more easily understandable by the general public, the Australian Department of Transportation and Regional Services (DOTARS) has developed the Transparent Noise Information Package (TNIP) [DOTARS, 2003]. The TNIP is a suite of software applications that uses output from the INM to produce such items as flight path movement and respite charts and measured N70 reports

AIR QUALITY AND MODELING OF POLLUTANT EMISSIONS

Aviation-related activities are sources of a variety of gaseous and particulate emissions with potential effects on global climate change and on local air quality. Within the context of airport operations, the effects of the aviation-related activities on local air pollution are of primary concern. Although the total contribution from aircraft to local air pollution in the vicinity of an airport is usually small compared to other non-aviation sources, its impact on the environment is not negligible.

Through its Committee on Aviation Environmental Protection (CAEP), the International Civil Aviation Organization (ICAO) has established standards for emissions certification of aircraft engines, expressed in terms of mass of emissions per unit of engine thrust, on nitrogen oxides (NO_x), carbon monoxide (CO), unburned hydrocarbons (HC) and smoke [CAEP, 2007]. Presently, there is no ICAO standard for aircraft engine Particulate Matter (PM). The ICAO emissions standards for aircraft engines have subsequently been adopted into domestic regulations by individual ICAO contracting states. For emissions certification purposes, ICAO has defined a specific reference Landing and Take-Off (LTO) cycle below a height of 915 m (3,000 ft) above ground level, in conjunction with its internationally agreed certification test, measurement procedures and limits. Aircraft pollutant emissions above this particular height are believed not to have any discernable impact on the air quality at ground level in the vicinity of the airport. The LTO cycle as defined by ICAO is a significantly simplified version of an actual operational flight cycle and is assumed to consist of only four modal phases, representing approach, taxi/idle, take-off and climb respectively.

In many developed countries, air quality is regulated through legislation by establishing standards on emission levels of various pollutants, and by defining procedures for achieving compliance with these standards. To quantify the impact of pollutant emissions computer databases and models have been developed. The two main strands of air quality assessment are (1) emissions inventories, to assess the total mass of emissions released into the environment resulting from all air traffic operations, ground service equipment, ground vehicular traffic and stationary sources during a specified period of time, and (2) dispersion modeling of pollution concentrations, to assess concentrations of various pollutants at specified locations in the vicinity of the airport for ambient atmospheric conditions. Aircraft emissions can be analyzed at two different levels. In some cases, an emission analysis remains restricted to an inventory only, while in other cases a dispersion study needs to be included as a second step. In the US, for example, the FAA requires the use of the Emissions and Dispersion Modelling System (EDMS) for conducting air quality analyses of aviation emission sources from proposed airport projects. The EDMS can be used to create an emissions inventory for any individual airport emission source, or combination of emission sources. For dispersion analyses, EDMS generates input files to be processed by its module AERMOD, which essentially implements a steady-state plume model for specified weather patterns [FAA OEE, 2007]. The EDMS includes emission indicators for the various airport sources. For example, it incorporates all aircraft engine emissions data contained ICAO Exhaust Emissions Data Bank, representing nearly two-thirds of EDMS's aircraft engine emissions data. The remaining aircraft engine emission data originate from other sources, including engine manufacturers.

In Europe, EUROCONTROL has developed the Advanced Emission Model (AEM), a system for estimating aviation fuel flow rates and emissions, including CO_2, H_2O, SO_x, NO_x, HC, CO, Benzene, Volatile Organic Compounds (VOC), Total Organic Gases (TOG), and PM [Jelinek, 2004]. The AEM uses several underlying system databases (aircraft, aircraft engines, fuel flow rates and emission indices) provided by external agencies, including the ICAO Engine Exhaust Emissions Data Bank.

The aircraft emission calculations conducted with the EDMS or the AEM are typically based on the LTO cycle below 3,000 ft, as defined by the ICAO Engine Certification. These calculations rely on the ICAO Engine Exhaust Emissions Data Bank to provide fuel flow rates (kg/s) and emission indices (g/kg of fuel) for a large number of engine types at four specific engine power settings (modes): 100% (takeoff), 85% (climb out), 30% (approach), and 7% (idle). For a single LTO cycle of an aircraft of type j,, the total emission Ei,j of pollutant i (grams) is given by:

$$E_{i,j} = \sum_{k} EI_{i,j,k} \cdot FF_{j,k} \cdot t_{j,k} \cdot n_i \quad ,$$

where:

$EI_{i,j,k}$ = emission index for pollutant i of aircraft type j operating in mode k (grams emitted per kg of fuel burned)

$FF_{j,k}$ = fuel flow rate of aircraft type j and mode k (kg/s)

$t_{j,k}$ = time in mode k for aircraft type j (seconds)

n_i = number of engines for aircraft type j

In the AEM, the ISA sea-level values of the ICAO emission indices are extrapolated to the operating altitudes and temperatures encountered throughout the aircraft flight profile, using a modified version of the CAEP-recommended

Boeing Method 2 [Jelinek, 2004]. Boeing's Method 2 also allows corrccting the ICAO aircraft engine certification values for installation effects and intermediate thrust levels. Figure 4 illustrates typical functions of emission indices of NO_x and CO versus engine power setting for a turbofan engine at different altitudes [Penner, 1999l]

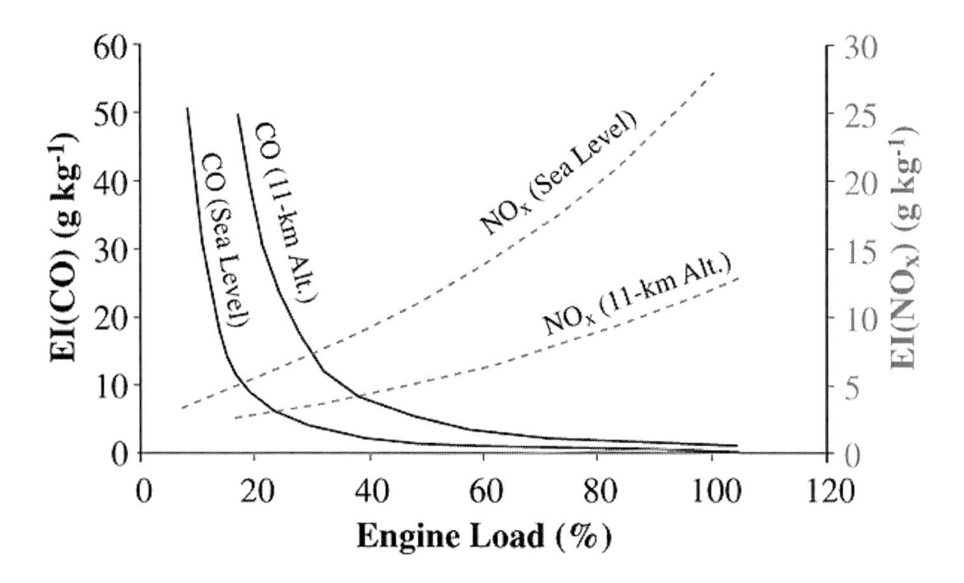

Figure 4. Emission indices of a turbofan engine versus engine load at two different altitudes; Source: [Penner, 1999].

THIRD PARTY RISK

An environmental concern that is gradually emerging on the political agenda in many developed nations is third party risk (also called external safety risk), or more explicitly, the risk of an aircraft crashing into a populated area in the vicinity of an airport. Due to the concentration of air traffic movements around an airport, the population living in its vicinity is (involuntary) exposed to a relatively high risk of aircraft accidents. Although the probability of an accident per flight is quite small, the actual local risk levels around an airport can still be significant, simply due to the large number of flight movements taking place into and out of the airport. Also the fact that crash rates during landing and take-off are high when compared to the en-route flight phase - some 70% of aircraft crashes happen close to the airport [Hale, 2002] - contributes to the situation that populated areas near the airport are exposed to a much larger external safety risk in comparison to the more remote residential areas.

Third party risk around airports is still a relatively new issue for airports, and at present there are very few airports where conducting a formal, quantitative risk assessment has become commonplace. Two countries where legislation exists to give guidance on what level of third party individual risk is acceptable, are the UK and The Netherlands. In both countries regulating third party risk primarily takes place in the context of land-use planning to prevent inappropriate development in risk zones.

To facilitate the development of risk zoning policies, methodologies to quantify the volume of risk and the distribution thereof have been developed in both the UK and The Netherlands. The risk assessment methodologies adopted in these two countries broadly follow a similar approach. In both countries it was concluded that the most appropriate metric for calculating third party risk around airports is individual risk. Individual risk is generally defined as the chance per year that an individual residing permanently at a particular location will be killed as a result of an aircraft impact. It is expressed in units per year. Based on the local individual risk calculations for the entire area around an airport, risk contours can be contructed by joining points on a geographical map which are subject to the same individual risk. The assessment of the individual risk is indicated in the diagram shown in Figure 5.

The method used to calculate individual risk at a particular point in the vicinity of an airport comprises three main steps. First, the probability of an aircraft accident in the vicinity of the airport must be determined, for each aircraft type class. This probability depends on the probability of an accident per aircraft movement (a landing or a take-off) and the number of movements carried out per year for each aircraft type class. The probability of an accident per movement, the accident rate, is determined from compiled historical data. The second step involves the determination of the distribution of crashes with respect to location. The probability of an accident close to the runways is higher than at larger distances from the airport. Also, the local probability of an accident is larger in the vicinity of arrival and departure routes. This dependence is represented in a crash location probability model, which, given that a crash has occurred, determines the probability of a certain location as a crash site. The accident location probability model is based on compiled historical data on crash locations. The last calculation step determines the size of the affected area and the proportion of people likely to be killed within this area, as a function of the aircraft parameters, impact parameters, and possibly terrain type.

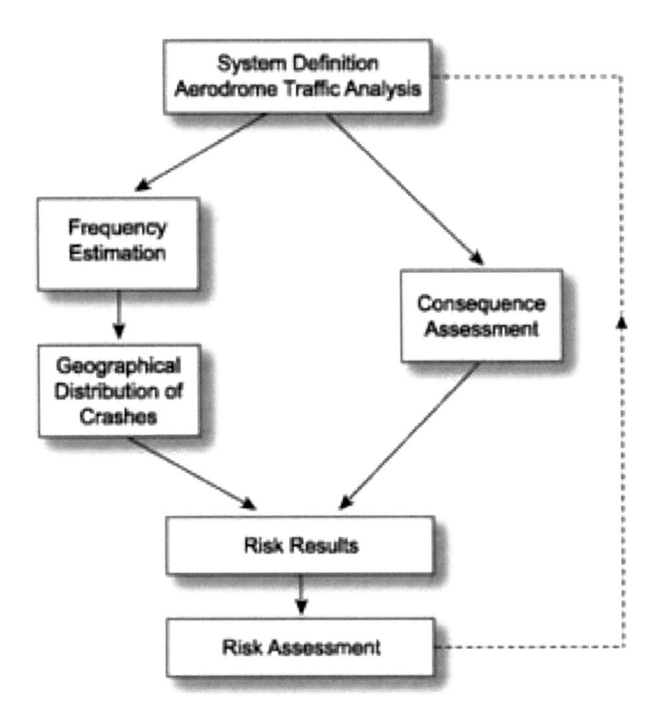

Figure 5. Diagram illustrating the three basic steps in the calculation of individual risk contours; Source: [EUROCONTROL, 2005].

Individual risk contours are used for zoning purposes at Schiphol airport in The Netherlands and at airports across the UK. In the UK, the Health and Safety Executive has recommended that third parties should not be exposed to risks over 10^{-4} per year. In addition, it has been recommended that the UK base so-called Public Safety Zones (PSZ) around airports on the 10^{-5} per annum individual risk contour. New or replacement development, or changes of use of existing buildings, are not permitted within Public Safety Zones. In the Netherlands, the Ministry of Housing, Spatial Planning and the Environment and the Ministry of Transport, Public Works and Water Management have jointly developed the external safety policy related to aviation, essentially setting construction limits or bans within certain areas with an increased risk. At present, the 5.10^{-5} per annum individual risk contour is used as the limit of the demolition zone. Homes within this zone must be demolished no later than 2015 [Hale, 2002].

Figure 6 shows a risk contour map for Schiphol airport in The Netherlands. The Individual risk levels, indicated by the 10^{-5}, 10^{-6} and 10^{-7} per annum contours, have been calculated using the NATS Third Party Risk Model, developed by the National Air Traffic Services (NATS) in the UK [Foot, 1997].

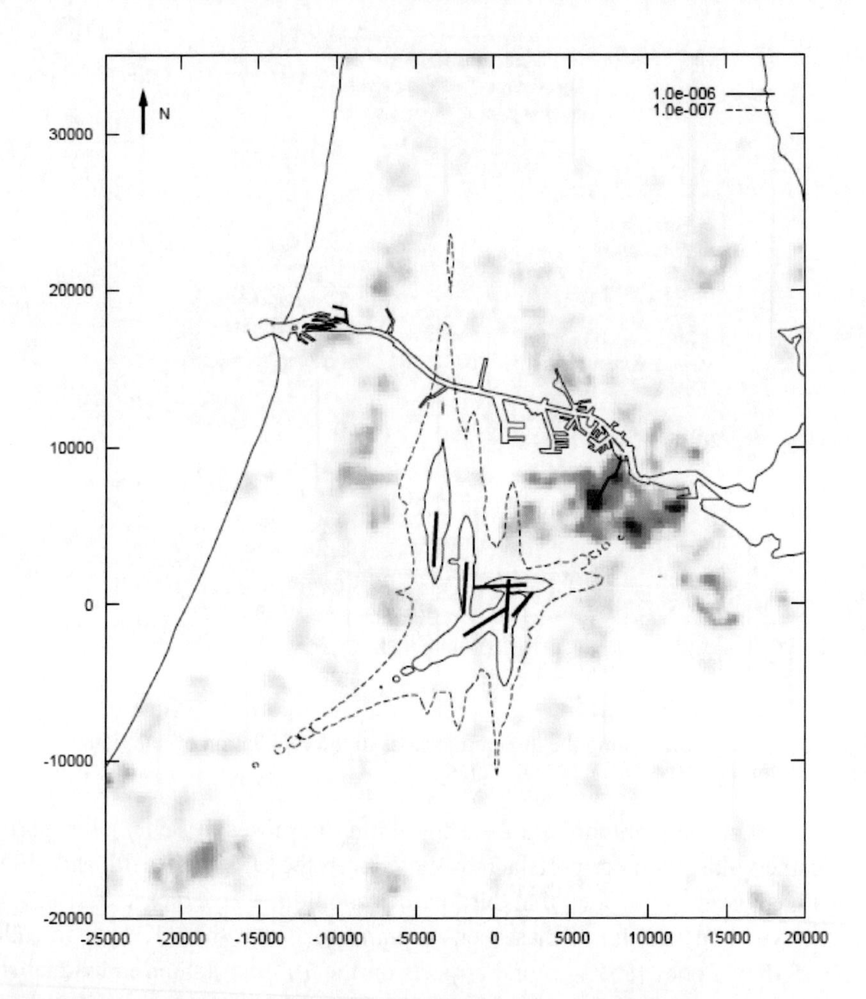

Figure 6. Individual risk contour map for Amsterdam airport Schiphol.

Chapter 2

INTEGRATED ENVIRONMENTAL MANAGEMENT CONCEPT

LEVELS OF AGGREGATION

To mitigate the negative environmental effects of aviation, a whole range of measures is available to airports close to sensitive communities. When focusing on these measures, different levels of aggregation can be identified. A possible categorization from lowest to highest is the following:

1. Source level: The aircraft (design) itself is the most basic level at which aviation related environmental impact reduction can take place. The main efforts of noise and emission reduction take place during the design stage of the aircraft or engine. Owing to progress in design knowledge and technology, new aircraft and engine designs typically outperform their predecessors. During the operational phase of the aircraft life cycle, modifications to improve the environmental performance are not very common, but do occur from time to time. Installing engine hush-kits to meet noise regulations would be an example.
2. Trajectory level: This level concerns the trajectory in 3D–space of a single aircraft. Trajectories can be separated into a horizontal and a vertical part. The horizontal part is referred to as the route or track, whereas the vertical part is called the (vertical) profile or the procedure. From an environmental point of view, it is preferable to optimize both elements of the trajectory simultaneously. However, procedure optimization is currently more of an international affair, aimed at the

development of standardized operating procedures. In contrast, route optimization is typically a local, airport specific issue.

3. Tactical-/operational level: This level concerns the actual day-to-day operation of the airport and the provision of air traffic control in the context of environmental performance. Air traffic controllers organize arrival and departure flows and provide guidance and separation of approaching, and departing aircraft in a TMA. Their main objective is to make optimal use of available runway capacity, and to achieve this they rely on tactical processes such as sequencing, radar vectoring and holding. At present, the involvement of ATC in the mitigation of environmental impact is rather limited. It is especially at this level where environmental objectives conflict with other areas of interest, like efficiency and safety.
4. Strategic level: Finally the strategic level encompasses all long-term and vision driven initiatives. This includes regulatory issues, such as land use planning and noise compatibility, and airport use and development. Sound insulation, property acquisition, other means of compensation and public involvement programs represent issues at the strategic level as well.

Although most environmental impact reduction efforts can be categorized into one of the four levels presented above, some projects actually span multiple levels. Recently, noise abatement procedure research projects are being focused on the trajectory as well as the tactical-/operational level [Optimal, 2007]. Interrelations between the levels can also be less intentional and even work counterproductive. For example, improvements in the design concerning the aerodynamic efficiency will reduce fuel burn and global emissions, especially during the cruise phase. If this improvement also improves the climb performance, the optimal departure trajectory may change as well. Typically, better climb performance also improves noise performance. However, for arrivals the situation is different. Reduced aerodynamic drag can make it harder to dissipate energy during the execution of a steep (continuous descent) approach, which is one of the favored tactics in approach noise reduction.

Integrated Environmental Management

In the ideal situation, all environmental impact mitigation efforts should be managed concurrently. When using such a form of integrated environmental management, it can be ensured that all actions taken to minimize the nuisance caused by aircraft noise and emissions will be consistent, complement each other, and make use of synergy benefits. At the same time, it helps avoiding that a decision at one level (partly) reduces the effectiveness of another measure at a different level.

However, different organizational bodies control the different levels of aggregation. From an airport management point of view, especially the source level is out of reach, as this level is mainly in control of engine and airframe manufacturers. The other three levels on the other hand can be influenced much more by the airport, especially when there is a close cooperation with the responsible air traffic control provider. Therefore, the Integrated Environmental Management Concept presented here will focus on the trajectory, tactical-/operational and the strategic level. Nevertheless, technological developments at the source level should be fully embraced and optimally exploited.

When looking at what has been achieved in the past decades at the three different levels, the tactical-/operational level seems to lag behind. Probably, the main reason for this is that the implementation of noise abatement procedures at the side of ATC authorities is not always straight forward, as it may interfere with requirements with respect to safety and efficiency. This can be observed when considering the current implementation of the Continuous Descent Approach (CDA). Considered in isolation (at the trajectory level,) the CDA procedure appears to be a very attractive mitigation measure; however at the tactical-/operational level implementation problems arise. As a result, the current trade-off for this procedure is either to accept a less than ideal continuous descent, or to accept a reduced arrival capacity [Davidson Reynolds, 2006; Kershaw, 2000].

Conflicting requirements are not the only problem that air traffic controllers face with respect to reducing community noise impact. Taking noise beneficial decisions can also be difficult simply because of a lack of noise-related information. Usually, controllers have access to 'static' information, like the preferred use of certain routes and runways. However, they are not provided with information on the continuously developing situation with respect to community noise exposure. This means for example that they cannot respond to developments in the noise exposure in the past or expected developments in the near future. Nor can they evaluate the environmental effects of a tactical or operational decision they are about to take, except from their own knowledge with respect to environmental issues.

Unfortunately, most efforts at the trajectory and the strategic level do need implementation at the tactical-/operational level. Eventually, this is the level where the actual execution of the flights takes place, resulting in noise and pollutant emissions. This means that without a means to support the initiatives at the tactical-/operational level, a significant part of potential environmental impact reduction simply cannot be brought to fruition. It should be seen as the connecting element in integrated environmental management. It is therefore, that especially this level will need a lot of attention in future research.

DECISION SUPPORT SYSTEM

A possible setup for the integrated environmental management concept would be in the form of automated decision support at the side of the air traffic controller, while making use of the concept of trajectory based navigation. Different studies into the future of ATC already foresee a transition from the current flight plan based navigation towards a trajectory based operations [JDPO, 2007]. This transition is expected to increase accuracy and detail of the aircraft trajectory in the three spatial dimensions. However, the increase in accuracy in the time dimension is important as well. The four-dimensional trajectories will give the air traffic controllers a far more accurate projection of which aircraft will be where at what time. As a result, they will be able to perform traffic situation predictions for periods further than just a few minutes ahead without the current problem of sharply degrading accuracy of the prediction.

In such a working environment, the tasks of the controllers can shift from a more controlling or operating tasks to a more supervisory, planning-oriented task. Together, the more accurate positioning and a more planning-oriented task of the air traffic controller will greatly improve the possibility to manage the allocation of environmental effects with respect to individual movements. This means that an already envisaged transition towards trajectory based navigation will provide a very good opportunity to combine environmental management with the traditional responsibilities of air traffic control. Especially if such a system would enable aircraft to fly individually customized and optimized trajectories, several advantages with respect to environmental performance can be identified.

When using customized trajectories, it becomes possible to take individual aircraft performance into consideration. Currently, procedures for arriving and departing traffic are designed such that at least the great majority of visiting aircraft should be able to adhere to the procedures under a wide range of weather and wind conditions. Basically, this means uniform design for the weakest link, possibly inhibiting better performing aircraft to exploit their capabilities. A situation where each aircraft would be flying its own customized and optimized trajectory can not only increase airport capacity, it also offers the opportunity to optimize the trajectory with respect to the environmental performance of the individual aircraft.

With further advances in aircraft noise modeling and computing power, customized trajectories also offer the opportunities to take actual weather conditions into consideration for trajectory optimization, compared to today's practice of using average conditions. This can decrease the discrepancies between computed optimal trajectories and the actual executed ones, making trajectory optimization more efficient. This example clearly shows the level of interaction between the trajectory and the tactical-/operational level that such a Decision Support system (DSS) would allow.

Of course, the DSS should also interact with environmental impact reduction efforts at the strategic level. As just an example, an innovative approach would be to relate noise proofing programs to noise zoning based on uniform indoor peak level maxima. Based on the applied noise insulation for a particular zone, this automatically results in outdoor peak levels for that zone. Now, if the actual noise performance of each aircraft is considered in the trajectory synthesis process, their trajectories can be determined such that the chosen restrictions will not be violated. The first effect is that this may help in achieving a very fair, easily explainable and cost effective acoustic insulation program. A second effect is that it in fact benefits airlines using aircraft with good noise performance, because it gives them more routing flexibility, which generally result in a more efficient routing.

The next three sections illustrate some aspects of our research efforts at the different levels of aggregation. The tools that are emerging from this research are expected to contribute to the planned DSS. This may either be in their current form or through an alternative implementation based on a similar concept. At the trajectory level, the tool NOISHHH is presented. It can perform environmental optimization with respect to the local situation on 3D or 4D trajectories, thus considering both route and procedure simultaneously. At the strategic level, a noise allocation tool, called SNAP is described. It calculates the ideal annual distribution of flights over the different runways of the airports, based on environmental and efficiency criteria specified by the user. Finally, we discuss the conceptual development of the central element in the integrated environmental management concept, a controller decision support tool at the tactical-/operational level.

Chapter 3

Optimization of Noise Abatement Trajectories

Advanced Noise Abatement Procedures

Most current noise abatement procedures are local adaptations of generic procedures, aimed at optimizing aircraft noise footprints [Erkelens, 2000]. Quite often, noise abatement procedures are designed to minimize the area impacted by high-intensity aircraft noise. Although the application of a procedure that is designed in this fashion may indeed result in a reduction in the area impacted by high-intensity noise, this typically comes at the expense of an increased area of low-intensity noise. Moreover, the exposed area criterion can hardly be viewed as a metric for the true noise impact, which obviously depends on the actual population distribution in the vicinity of an airport.

An important constraint in the application of noise abatement procedures is that the airport throughput capacity during peak hours should not reduce. In other words, improvements with respect to noise impact should not come at the expense of a reduction in throughput capacity. Whilst current noise abatement arrival and departure procedures consider throughput efficiency to some extent, they do not generally consider emissions or fuel burn.

Although weather - notably wind and temperature (variations) - has a significant effect on aircraft performance and noise propagation, and, consequently, on the noise impact of aircraft operations, current noise abatement procedures are not adaptive to weather conditions, but generally rely on a single static design that corresponds to "average day" weather conditions.

To replace current single-purpose, static noise abatement procedures, we are currently directing research efforts towards the development of advanced environmental procedures that dynamically adapt to changing needs and that maximize benefits based on an integrated assessment of multiple relevant factors, including noise, fuel burn, emissions and throughput efficiency. To facilitate the design of such flexible-geometry noise abatement procedures, a tool called NOISHHH is being developed at the Delft University of Technology [Visser, 2001;Visser, 2003;Visser, 2005]. This innovative tool combines a noise model, a dose-response relationship, en emission inventory model, a geographic information system, and a dynamic trajectory optimization algorithm. The NOISHHH tool generates routings and flight-paths for both arrivals and departures that minimize the environmental impact in the residential communities surrounding the airport, while satisfying all imposed operational and safety constraints It must be noted that the NOISHHH tool only considers single event flyovers and does not assess the community noise impact based on an aggregation over all aircraft operations for a particular time period.

In the next subsections, the design and characteristics of the NOISHHH tool are highlighted, and the potential environmental benefits of the arrival and departure trajectories optimized with NOISHHH are illustrated in examples based on an approach from the North to runway 06 and a departure to the North from runway 24 of Schiphol airport in the Netherlands. The presented examples rely on a model of the Boeing 737-300 twinjet transport. A brief description of work in progress and anticipated future work is provided as well.

ENVIRONMENTAL TRADE-OFF FRAMEWORK

To facilitate informed decision making on environmental trade-offs (e.g., between noise and emissions) in the synthesis of advanced noise abatement trajectories, the NOISHHH tool concept has been principally based on a multi-objective optimization framework. The NOISHHH tool features a variety of (environmental) performance criteria, including, gaseous emissions, fuel burn, transit time and noise exposure. Typically, these different criteria are not compatible; the decision variables that optimize one objective may be far from optimal for the others. Improvement with respect to one particular criterion is often achieved at the expense of one or more of the other noise criteria. To permit a trade-off, NOISHHH considers a "composite" performance index that essentially consists of a weighted combination of the various criteria

The emissions inventory currently implemented in NOISHHH includes three pollutants, viz. nitrogen oxides (NO_x), carbon monoxide (CO), and unburned hydrocarbons (HC). Moreover, there are several pollutants that are directly proportional to fuel burn, including sulphur oxides (SO_x). The emission performance criteria implemented in NOISHHH correspond to the mass of each of these pollutants emitted below 3,000 ft AGL.

The NOISHHH trajectory-synthesis tool features a variety of noise performance criteria. Some of these noise criteria are generic in nature, e.g., a criterion that is based on the total area enclosed within a specified noise level contour, while others are site-specific in the sense that they depend on the density and distribution of the population in the vicinity of a particular airport [Visser, 2005]. A typical example of a site-specific noise criterion is the population count within a specified noise level contour. In the numerical examples presented in the Section, the noise impact assessment remains restricted to the site-specific criterion *Awakenings,* which represents the number of people within the exposed community that is expected to awake due to a single event nighttime flyover.

A performance criterion which is consistently included as a secondary criterion in every noise optimization study conducted with NOISHHH is *Fuel,* the fuel-consumed during an arrival or departure trajectory. Including fuel-consumed in the composite performance index is of particular importance to mitigate the occurrence of "exotic" horizontal arrival or departure routings and extremely large transit times that may result from bypassing noise sensitive areas.

The transit time for a departure or an arrival flight can be included as a criterion in the composite performance index. However, it is also possible to explicitly include transit-time constraints in the trajectory optimization formulation. This latter feature is particularly important in the context of the development of 4D noise abatement procedures.

The optimizations of the nighttime noise abatement trajectories presented in the numerical examples have been based on the following composite performance index:

$$J = Fuel + K \cdot Awakenings \quad ,$$

where $K\,(\geq 0)$ is a user-selected weighting factor.

Physical Modeling

To evaluate aircraft flyover noise, a model has been developed that essentially implements the basic methodology employed within the Integrated Noise Model INM – described in Subsection II.C - to compute sound exposure levels at specified observer locations. The observer locations are arranged in the form of a rectangular grid of points surrounding the residential areas in the vicinity of the approach or departure path. The size and mesh of the grid have a significant impact on the computational burden and must therefore be chosen judiciously. Due to the fairly long glideslope approach, the noise sensitive area in an arrival scenario is typically larger than in a corresponding departure scenario, and consequently, the grid area size selected in an arrival study is usually larger in comparison to a departure study. Grid area sizes that have been adopted in example studies typically range from 14x25 km^2 to 45x55 km^2, with a standard grid cell size of 1x1 km^2.

The specification of an awakening-related performance index requires knowledge of the relationship between aircraft noise and sleep disturbance. In NOISHHH the dose-response relationship as proposed by FICAN in 1997 has been implemented. This particular relationship, discussed in Subsection II.B, provides the percentage of the exposed population expected to be awakened (%Awakenings) as a function of the exposure to single event noise levels expressed in terms of sound exposure level occurring indoors. In NOISHHH the indoor sound level is approximated by lowering the outdoor level as predicted by INM by 20.5 dB(A), which is a value that represents the average sound absorption for a home. By combining the %Awakenings results with the actual population density distribution in the noise exposed residential communities, the absolute number of people likely to awake due to a flyover can be determined.

The emissions inventory implemented in NOISHHH is based on the ICAO Engine Exhaust Emissions Data Bank (see Subsection II.D). The ICAO databank provides a compilation of aircraft engine emission data measured at four thrust levels: 100% (takeoff), 85% (climb out), 30% (approach) and 7% (idle) of maximum thrust available for take-off under normal operating conditions at ISA sea level static conditions. In NOISHHH the emissions of HC, CO and NO_x for each thrust setting along a departure or arrival trajectory are calculated through a simple linear interpolation between the above tabulated emission data. The SO_2 emission level is directly proportional to the calculated fuel burn and is estimated using the emission coefficient of 0.84 grams per kilogram of fuel burnt [EEC, 2006].

The flight-path computation methodology implemented in the INM package has not been adopted in the NOISHHH development. Instead, a slightly simplified point-mass model based on [Visser, 2001] has been used to represent the aircraft dynamics. The optimal trajectory calculations are based on the so-called intermediate model. The underlying assumption for the intermediate model is equilibrium of forces normal to the flight path. The implication of the simplifying assumption is that the aerodynamic drag is slightly underestimated in the sense that it is now evaluated as if the aircraft performs a quasi-linear flight. This particular model has been used extensively in studies concerning fuel-optimal trajectories, where it demonstrated an acceptable level of accuracy for a typical commercial aircraft [Visser, 2001].

In the numerical examples, performance data (including fuel flow characteristics) pertaining to a Boeing 737-300 aircraft is used. The data set comprises separate drag polars for each flap setting and for the configurations with the undercarriage extended or retracted. At present, all calculations are performed using the assumption of standard atmospheric conditions, and no wind present.

TRAJECTORY OPTIMIZATION APPROACH

The numerical optimization method employed in this study to solve the dynamic trajectory optimization problem is the direct optimization technique of collocation with nonlinear programming (NLP). The collocation method essentially transforms an optimal control problem into a NLP formulation by discretizing the trajectory dynamics [Hargraves, 1997]. To this end, the time interval of an optimal trajectory solution is divided into a number of subintervals. The individual time points delimiting the subintervals are called nodes. The values of the states and the controls at the nodes are then treated as a set of NLP variables. The system differential equations are discretized and transformed into algebraic equations (implicit integration). The path and control constraints imposed in the original optimal control problem are treated as algebraic inequalities in the NLP formulation.

A major advantage of using a direct optimization approach is that, unlike indirect (variational) methods, there is generally little difficulty in imposing constraints along the flight path. However, for the problem at hand the main advantage of this approach lies in the fact that the collocation discretization is fully compatible with the discretization (segmentation) approach taken in the INM model.

To solve the described optimal control problem, a software package called EZopt has been used [Visser, 2001]. The simple variant of the collocation method implemented in this package results in piecewise constant control histories and piecewise linear state histories. EZopt proved to be ideally suited for this problem, especially since it turned out to be quite capable in dealing with so-called multi-phase trajectory optimization problems. Occasionally, discontinuities may occur in the dynamic system equations describing the motion of the aircraft. Such discontinuities tend to result in significant problems in the optimization process. By resorting to a multi-phase formulation these problems can be overcome. A multi-phase formulation allows discontinuities at phase transitions, as well as the implementation of different dynamical systems for different phases. The multi-phase feature of EZopt is of vital importance for the noise optimization procedure proposed herein, since system discontinuities do indeed occur due to changes in aircraft configuration (i.e., flap setting and undercarriage position) and thrust rating.

As mentioned earlier, the optimal control problem is solved by discretizing the time interval in smaller subintervals, to approximate the state and control variables. Generally, the accuracy of the numerical solution of the problem improves with an increasing number of subintervals, or equivalently, the number of grid points. On the other hand, increasing the number of grid points requires a larger computational effort, hence forcing a compromise between desired accuracy and computational burden. However, in the present optimal noise abatement study, the flight path segmentation has been primarily based on the requirements associated with the computational methodology employed in the INM model.

NUMERICAL EXAMPLES OF ENVIRONMENTALLY OPTIMIZED DEPARTURE TRAJECTORIES

The NOISHHH tool has been used to generate optimized arrival and departure trajectories in a large number of numerical experiments. To illustrate a typical departure scenario, the so-called "Spijkerboor" departure at Schiphol airport has been selected, basically because it represents a fairly noise-sensitive route. As illustrated in Figure 7, the Spijkerboor departure involves a right turn shortly after lifting off from runway 24. A close inspection of Figure7 shows that an aircraft has to maneuver between the two residential areas, Hoofddorp (57,000 inhabitants) and Nieuw-Vennep (15,000 inhabitants). The right turn is followed by an interception of radial 213 of the Spijkerboor VOR. This particular route brings an aircraft fairly close to the densely populated Haarlem agglomeration (about 175,000 inhabitants). Moreover, on the other side of the flight track, Zwanenburg (8,000 inhabitants) is exposed to noise to a significant extent.

For a realistic representation of an optimized departure trajectory from runway 24 to the Spijkerboor VOR, the actual flight path is split up into four sequential phases. In the first phase, the aircraft flies a straight flight path starting at 122 m (400 ft) altitude and 90 m/s indicated airspeed (IAS), executed with a maximum take-off thrust rating and flap setting 1. The first phase terminates when the indicated airspeed reaches the flaps retraction speed (100 m/s). The second phase is similar to the first phase except that it is now flown in a clean configuration. The second phase terminates when an altitude of 457 m (1,500 ft) is reached and the maximum thrust rating is reduced to the maximum climb setting. In the third and subsequent phases turning is permitted. Also, a speed restriction is enforced in the third phase. The third phase ends when an altitude of 3,048 m (10,000 ft) is reached. The final (fourth) phase is similar to the third phase except that the speed restriction is no longer applicable. The final phase terminates at an altitude of 4,500 m. In a multi-phase formulation so-called staging (or phasing) conditions need to be included. Staging conditions are constraints that specify how the state at the end of a particular phase corresponds to the initial state in a subsequent phase. In the present formulation, the staging conditions are quite simple in the sense that the initial state of a particular phase is directly and fully connected to the terminal state of the preceding phase.

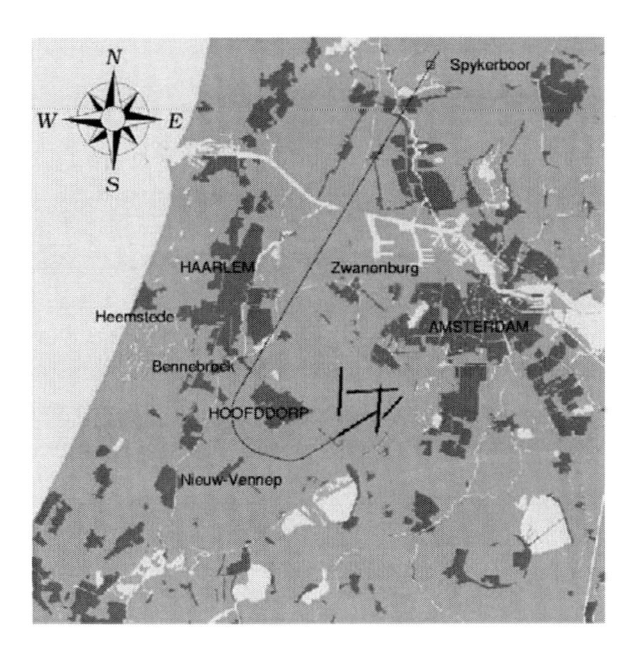

Figure 7. Spijkerboor departure at runway 24 of Schiphol airport.

The equations of motion are written in a coordinate frame fixed in space that has its origin at the grid point closest to the initial point of the departure trajectory. This initial point is located on the runway centerline extension, some 2,800 m from the threshold of runway 24. To ensure a flight of sufficient long duration, the end point of the flight trajectory is not located at the Spijkerboor VOR, but rather at some point well beyond it. All components of the final state vector have been specified, primarily to ensure that noise abatement does not result in unrealistic behavior in the terminal flight phase. The specified final altitude is 4,500 m and the required terminal speed is 275 kts IAS. These values have been selected based on earlier minimum-fuel studies. The final value for the heading angle is specified as 33°, to allow the final stage of the flight to conform with the existing Spijkerboor departure (i.e., flying along radial 213 inbound to Spijkerboor VOR). To enable thrust cutbacks, the flight time has not been specified in the examined scenarios.

A variety of constraints needs to be imposed in any given scenario. These constraints arise as a result of existing operational requirements, and also ensure that the trajectories generated by the optimization tool remain practically feasible and more or less resemble the flight paths that are currently flown. The usual constraints include a limit for the aerodynamic roll-angle (in this example a value of 25° is assumed), and operational speed limitations. Particularly notable is the ATC imposed speed constraint of 250 knots IAS on climb-out paths, below 10,000 ft altitude. For the existing Spijkerboor departure additional speed constraints are in effect, in the sense that during the right turn the indicated airspeed should not exceed 220 kts IAS. The latter constraint has also been included in the present trajectory optimization formulation. In order to obtain realistic results additional path constraints need to be included to inhibit (local) loss of speed (IAS) and altitude (h) during a departure flight. Some care has to be taken in the formulation of the thrust constraints. The application of thrust cutback should be compatible with aircraft performance to the extent that the engine-out climb gradient can be maintained after engine failure and subsequent thrust restoration. The application of reduced thrust should thus depend on the aircraft gross weight. Performance-related thrust constraints have, as yet, not been implemented in NOISHHH. The simple approach taken here is to enforce a constraint on the relative thrust:

$$0.6 \le \frac{Thrust}{Thrust_{max}} \le 1$$

In other words, the thrust level can not be reduced below 60% of the maximum value. A more detailed study involving the influence of more realistic thrust constraints on the optimal noise abatement procedures is envisaged for the near future. Finally some geometric constraints have been introduced, that are needed to insure that noise optimized tracks exit the noise-observer grid area at an appropriate location. The geometric constraints can be removed if the grid is enlarged to the extent that the entire trajectory is contained within it.

In order to demonstrate the optimization capabilities of NOISHHH, we present an environmental performance trade-off between arrival trajectories that have been optimized for a range of different weighting factors *K* in the composite performance index (see Subsection IV.B). When the value of the weighting factor K in increased from zero (minimum-fuel problem), the emphasis in the optimization process shifts towards minimizing the expected number of awakenings. Figure 8 shows the fuel-burn and awakenings (noise) performances of the optimized departure trajectories for a range of values of the weighting factor *K*. As expected, the number of awakenings sharply decreases as the weighting factor *K* increases. It is noted that for values of the weighting factor *K* in excess of 0.01, the optimization algorithm fails to converge. So even the noise-optimized solution that provides (relatively) the best noise performance in Figure 8 (i.e., the solution corresponding to K = 0.01) is to a large extent still shaped by fuel considerations.

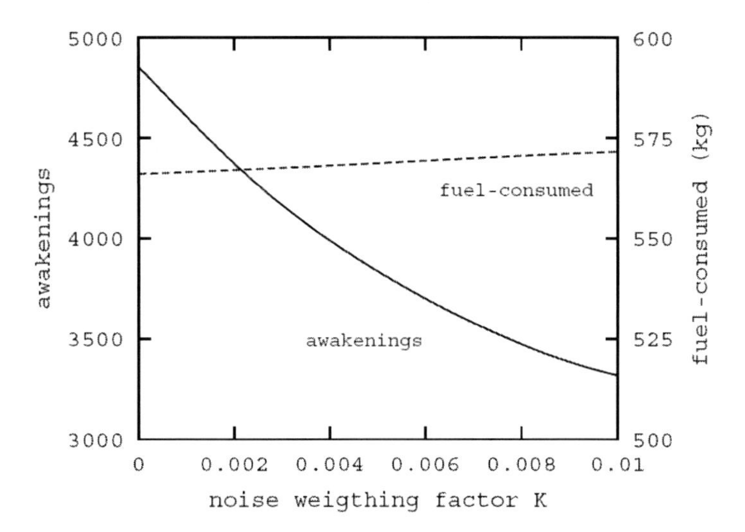

Figure 8. A comparison of optimal performances for various values of the weighting factor *K*.

In Figure 9(a) some key results pertaining to the minimum-fuel solution (*K* = 0) are shown. Similarly, Figure 9(b) depicts results for a noise-optimized trajectory ($K = 0.01$). Figures 9(a) and 9(b) show the aircraft departure flight paths in three-dimensional space, along with projections of the flight paths onto the horizontal plane (ground tracks). Note that only the initial part of an overall flight trajectory is shown. In addition to the flight path, also the (14x25 km^2) noise grid and a contour plot associated with a key metric is shown in each of the four graphs in Figure 9(a) and in Figure 9(b), respectively. The four selected metrics are, respectively, outdoor Sound Exposure Level (SEL), the percentage of people expected to awake (per square kilometer), the population density distribution, and the number of people awakened as a result of the flyover.

To examine the characteristics of the optimized trajectories, it is of interest to compare the minimum-fuel and noise-optimized trajectories presented in Figures 9(a) and 9(b), respectively. First, it is readily clear that for the fuel-optimal trajectory the 80-dB(A) noise contour is somewhat larger, whereas the 70-dB(A) noise contour is actually substantially smaller than for the noise-optimized trajectory. As a matter of fact, the 70-dB(A) contour is completely contained within the noise grid only for the minimum-fuel case. Apparently, improvements in noise performance in the proximity of the departure runway come to a certain extent at the expense of the noise performance in more distant regions. Not surprisingly, the percent awakening contours are very similar in shape to the SEL contours. As has been pointed out before, the true noise impact heavily depends on the actual population distribution. A close inspection of the results shows that, especially in the vicinity of the city of Haarlem, the noise impact can be significantly reduced in a noise-optimized trajectory. However, in the relatively small community of Zwanenburg, the noise exposure is actually somewhat increased. This is because the noise-optimized trajectory now directly overflies Zwanenburg. Nevertheless, the overall improvements in noise impact that can be obtained are fairly impressive for the considered scenario. Indeed, the number of people that is expected to awake due to noise impact reduces from 5,042 for the fuel-optimal trajectory to 3,312 for the noise-optimized trajectory, a reduction of nearly 35%. The total population living within the area defined by the noise grid is 371,635. Consequently, the reduction in awakenings in relative terms is from approximately 1.4% to about 0.9%. With respect to fuel consumption, the differences between the minimum-fuel and noise-optimized trajectories are rather modest. The noise-optimized trajectory requires only about 1% more fuel.

To illustrate how the improvements in noise performance are brought about, some more detailed trajectory optimization results are shown in Figure 10. The first two phases (flown with maximum takeoff thrust) are essentially the same for the two trajectories. In the first phase, the aircraft accelerates horizontally to the flap retraction speed. In the second phase, the aircraft climbs at a constant airspeed up to 457m (1,500 ft) above airport elevation. In the subsequent third phase some major differences in controlling the flight path arise. Recall that in this phase a full-bank right turn takes place. Unlike in the noise-optimized trajectory, the climb at constant airspeed is initially maintained in the third phase of the minimum-fuel flight path. At an altitude of about 720 m, the minimum-fuel trajectory features a horizontal transition segment during which the aircraft accelerates to the 220-kts IAS limit. Such a low-altitude horizontal turn is actually a fairly typical feature for minimum-fuel trajectories. It is evident, however, that from a noise perspective a horizontal turn at low altitude, executed at full thrust, is less desirable.

In the noise-optimized trajectory, the aircraft transitions much earlier to the 220-kts IAS limit. To this end, the climb rate is reduced, and full thrust is maintained. As expected, no low-altitude horizontal flight segment is present in the noise-optimal solution. A thrust cutback takes place when the aircraft overflies the west side of Hoofddorp. The thrust cutback is actually the main instrument to reduce the noise impact in this area. Also the Haarlem agglomeration benefits from the thrust cutback. Moreover, the noise impact in the two latter cities is favorably influenced by the slightly extended right turn, followed by a modest left turn. This particular behavior allows the noise impact to be shifted from densely populated city areas to the more rural regions.

When the value of the weighting parameter K is in the composite performance index is increased, the extent of the thrust cutback is increased. Clearly, there is a limitation to the extent of the thrust cutback that can be sustained for any given set of boundary conditions. As mentioned earlier, in the present scenario it turns out that the weighting factor K can not be taken significantly larger than 0.01.

In view of the fact that the minimum fuel-solution exhibits a rather undesirable noise behavior, we have also explored an alternative optimization problem formulation, in which the composition of the performance was changed. More specifically, *Fuel* was replaced as the baseline performance criterion in the composite index by a penalty term related to deviations from a defined reference flight path [Wijnen, 2003]. In this study, the current Spijkerboor Standard Instrument Departure (ICAO-A take-off procedure) was used to generate the reference flight path. Any deviation from the reference flight path leads to an increase in the penalty term in the composite performance index. As a consequence, the optimization process will only result in deviations from the reference solution if this is more than offset by a reduction in the number of expected awakenings. The advantage of this approach is that it produces solutions than remain close to the current practice. For large values of the weighting parameter *K*, the approach leads to solutions that give about the same number of awakenings as found when fuel is part of the composite performance index.

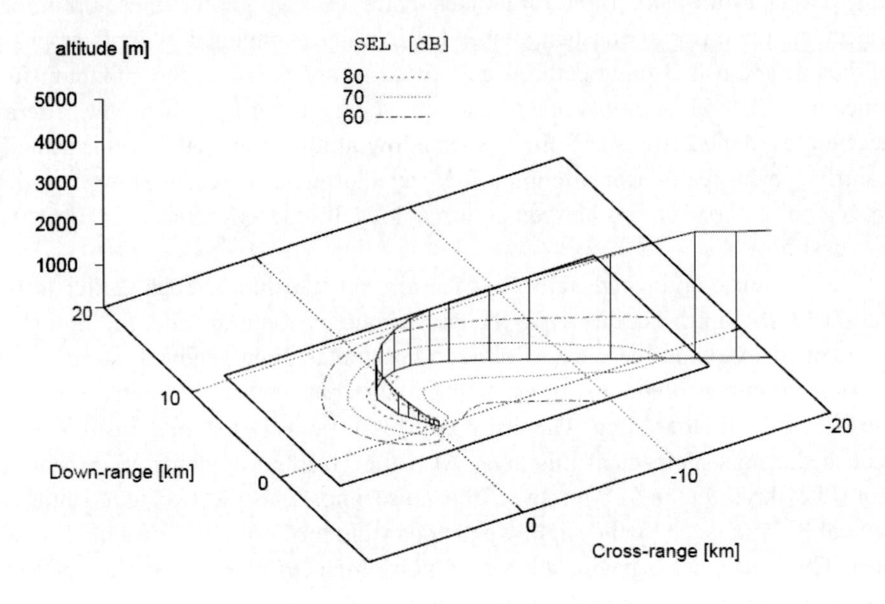

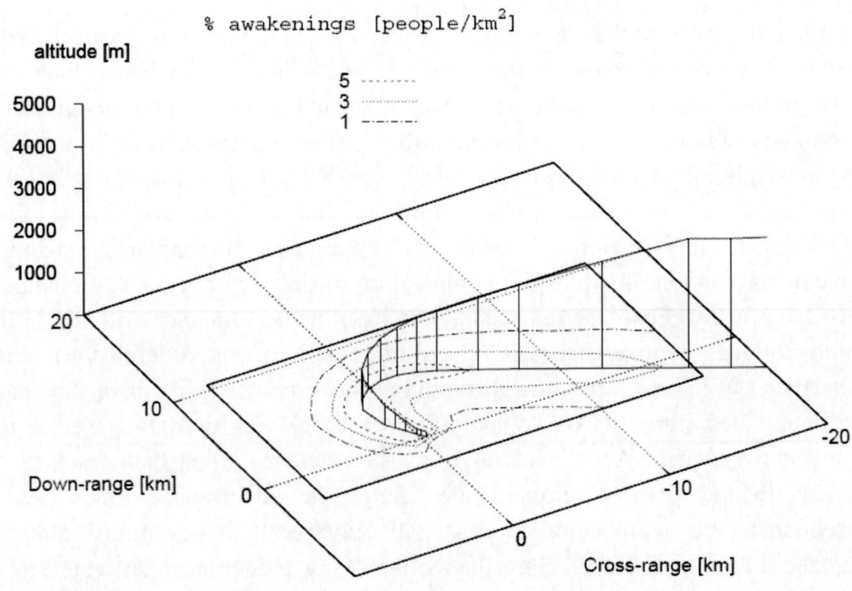

Figure 9a. (Continued)

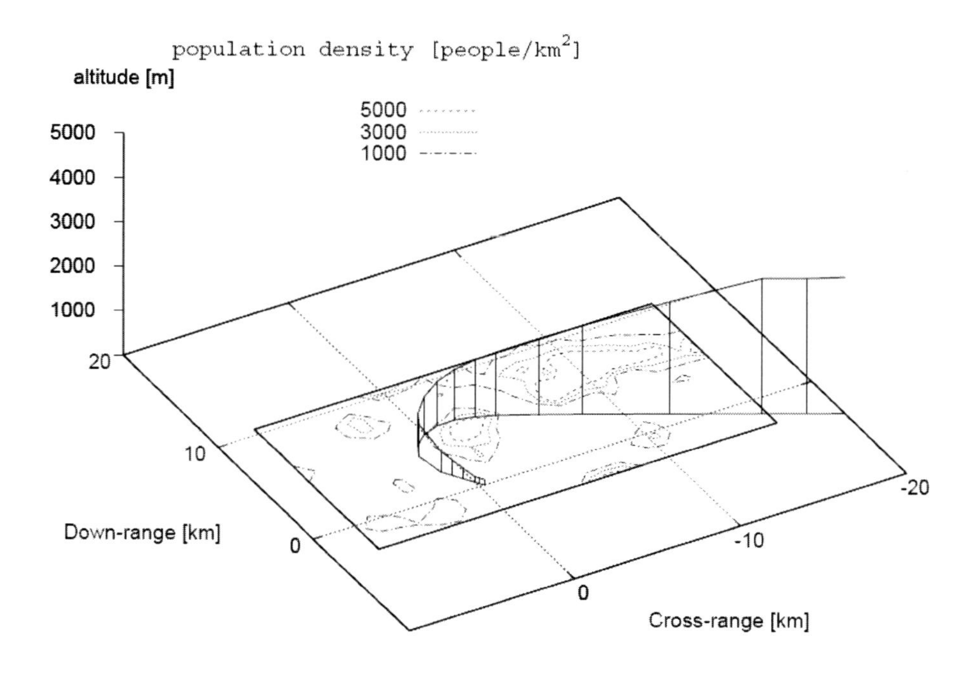

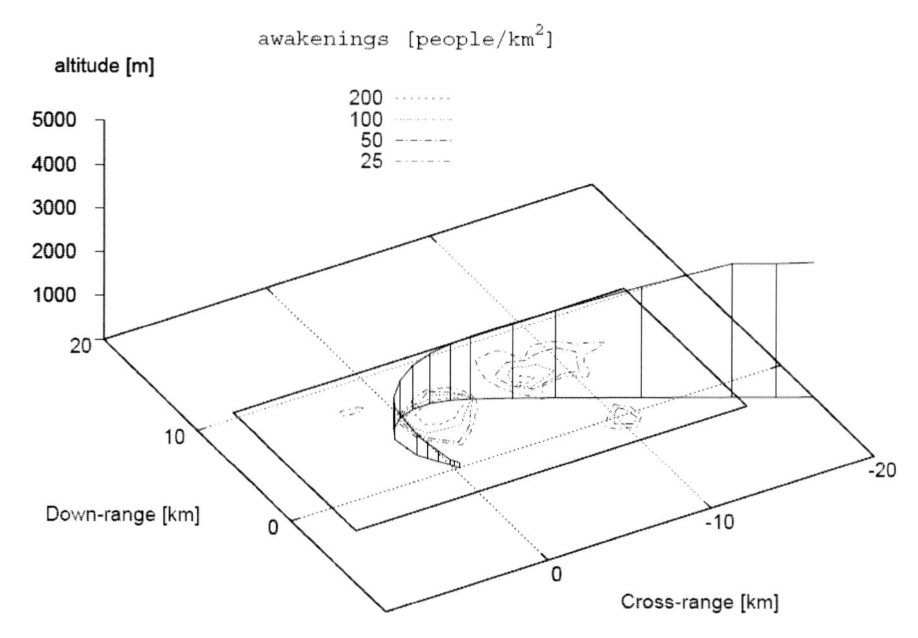

Figure 9a. Minimum-fuel departure trajectories.

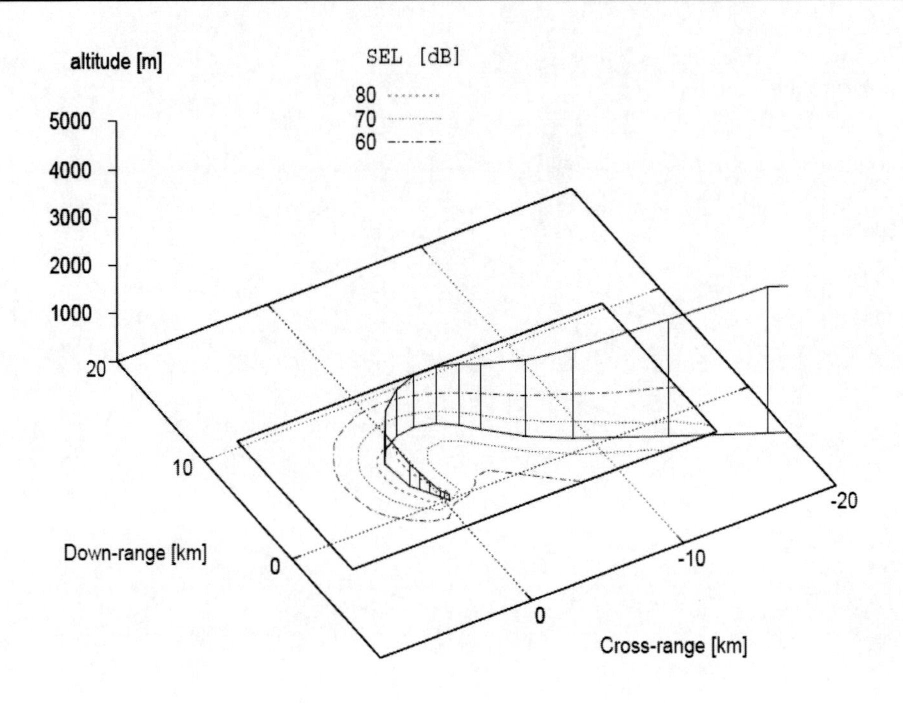

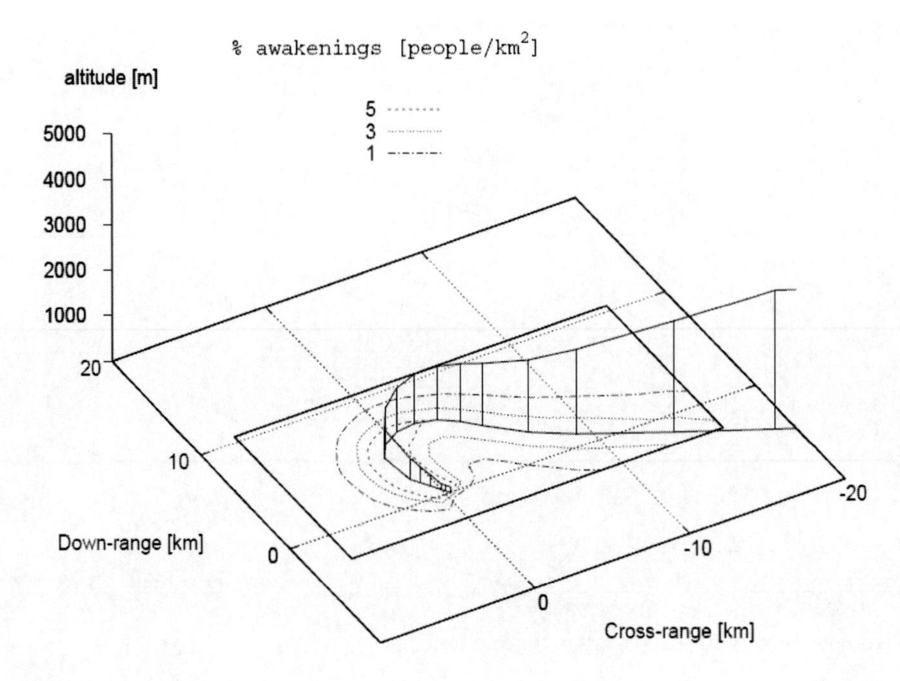

Figure 9b. (Continued)

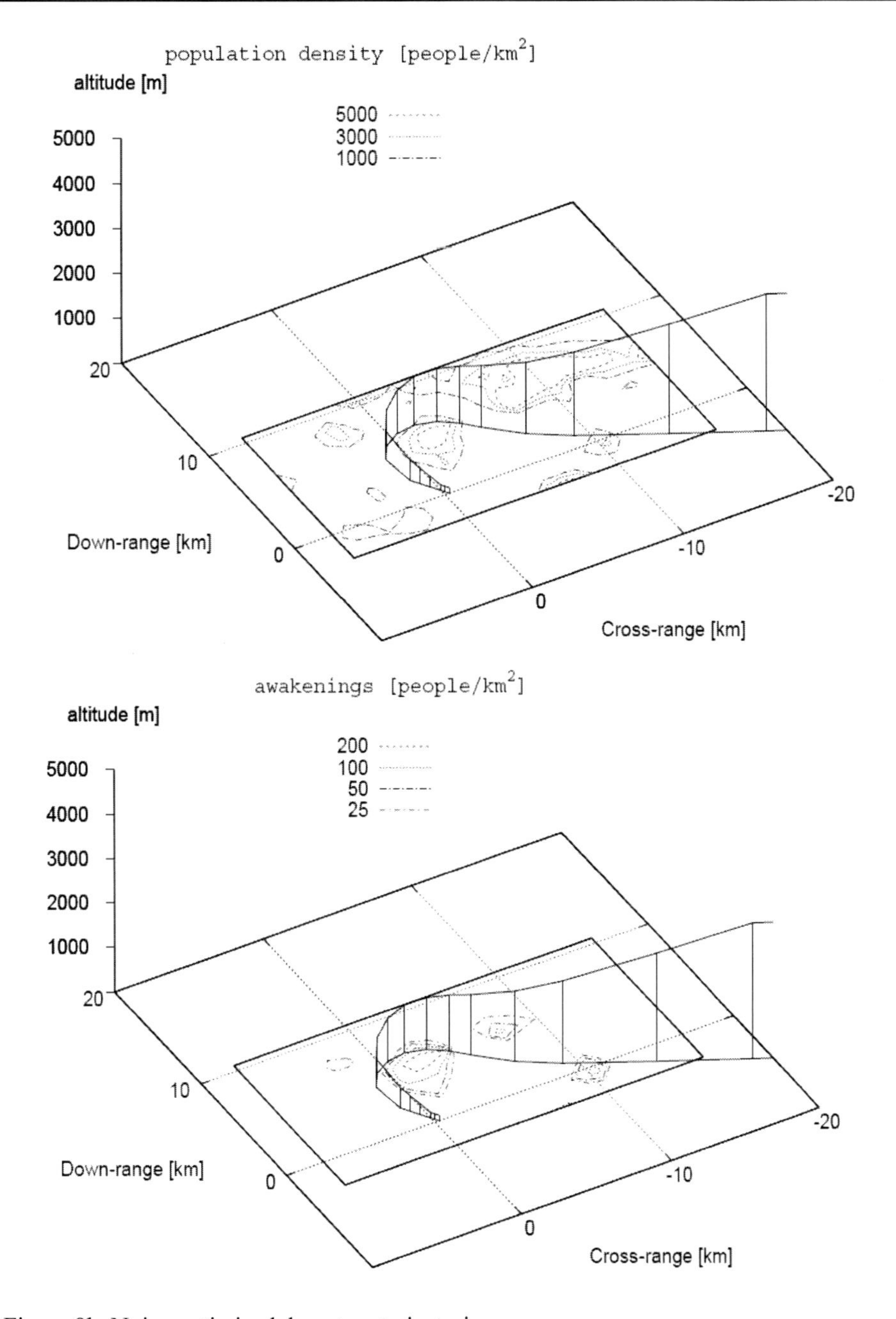

Figure 9b. Noise-optimized departure trajectories.

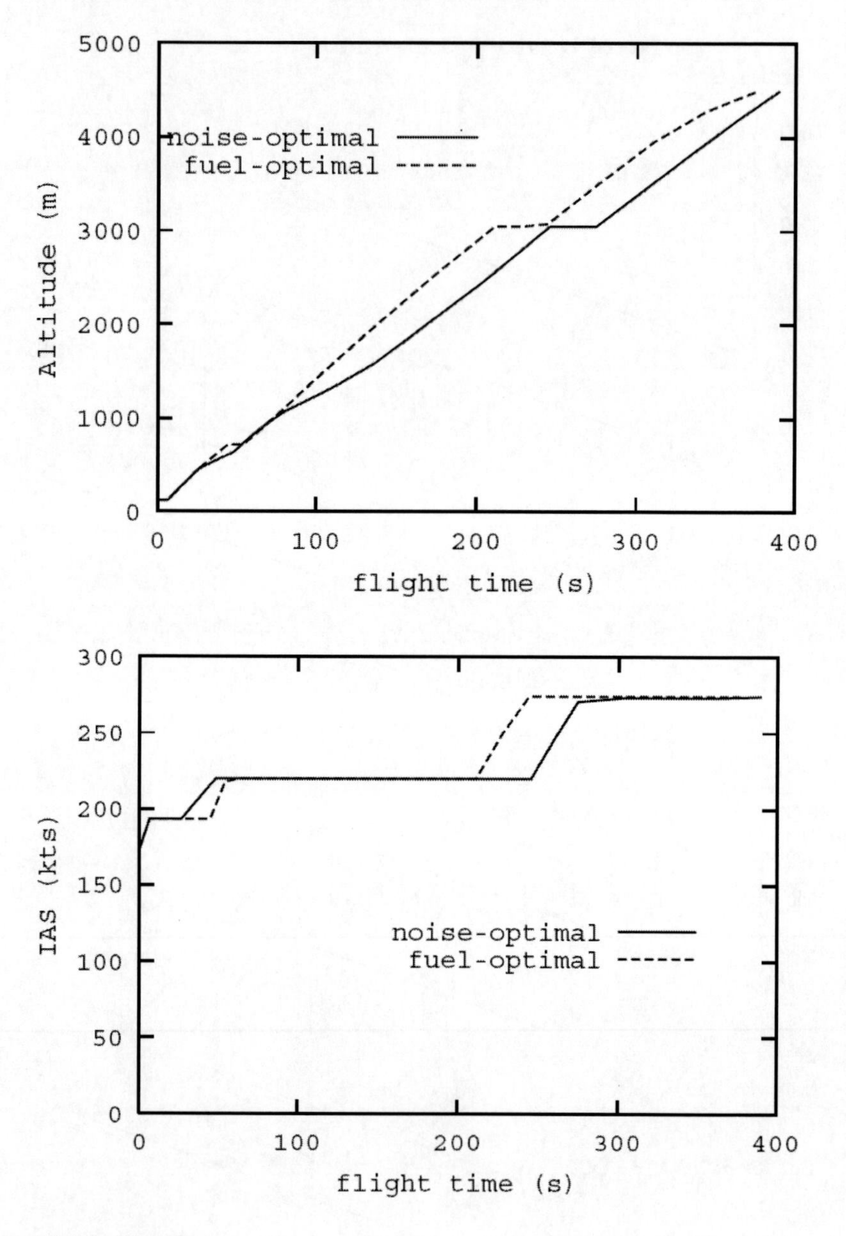

Figure 10. (Continued)

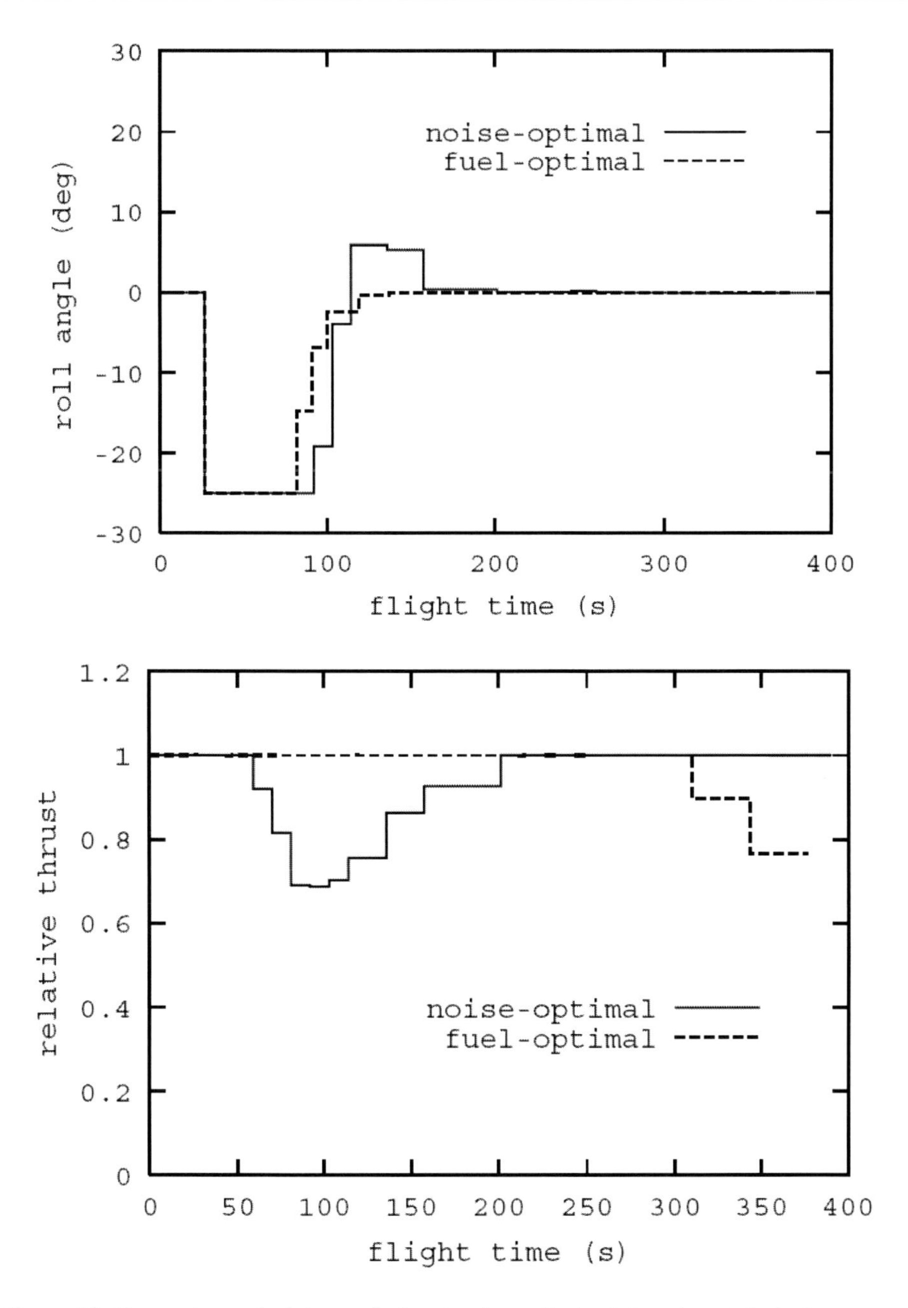

Figure 10. Comparison of minimum-fuel and noise-optimized departure trajectories.

AIR TRAFFIC MANAGEMENT OF ARRIVAL FLIGHTS

In current air traffic control practice, the desired separation between aircraft arriving in the terminal area is typically maintained through vectoring. In the vectoring process, air traffic controllers issue speed and heading instructions to pilots to stretch or shorten the flight path, with the aim to increase or decrease their separation from the aircraft in front or behind. Since vectoring instructions are most frequently issued when aircraft fly at low altitude, this way of working tends to result in a significant noise impact.

Approach flight path parameters that typically influence the noise footprint include the vertical (altitude/speed) profile and aircraft configuration (flap setting, undercarriage position). Navigation precision is also of paramount importance in this respect. The limited navigation precision provided by current onboard and ground systems in combination with the extensive use of radar vectoring results in significant average track dispersion and, consequently, in an extended area exposed to aircraft noise. It is expected that future guidance and navigation aids (e.g. Global Navigation Satellite System, Microwave Landing System) are likely to offer significant improvements in accuracy and coverage. When an approach flight path can be flown with high precision, track deviations can be kept small and, as a consequence, horizontal arrival routings can be shaped such that overflying noise sensitive areas can be minimized.

The advantage of a conventional approach is that a controller has full flexibility in shaping the lateral and vertical flight and speed profile, enabling him/her to increase runway arrival capacity through precise inter-aircraft spacing [Visser, 2006]. When RNAV noise abatement approaches are applied, controllers do not have such flexibility. This type of approach involves a pre-defined lateral trajectory, connecting a series of waypoints, along with altitude and speed targets in the vertical plane. Due to differences in e.g., aircraft performance characteristics, or FMS behavior, variations in speed profiles among aircraft may occur which hamper the controller in making accurate trajectory predictions. As a consequence, fairly large separation intervals in the arrival sequence are required, with obvious implications for runway throughput capacity

The uncertainty of an aircraft's future flight path could be significantly reduced, when airborne and ground-based systems would be able to share aircraft trajectory intent information. Whilst detailed trajectory intent information is available to the pilot through the FMS, this information is not readily available to the controller with current ATC technologies. To realize the huge noise impact reduction benefits of RNAV noise abatement operations in high density terminal airspace, a shift is required towards a new operational paradigm that relies on advanced technology to enable the sharing of intent information. The future operational concept that we envision seeks to address this issue by managing flights through the use of four-dimension (4D) trajectories, which accurately specify current and future aircraft position (latitude, longitude, and altitude) and time along specified points in the flight paths. A major benefit of 4D arrival trajectory management is the ability to check trajectories for potential conflict and to increase overall throughput efficiency by providing optimally spaced landing times. In the proposed concept it is implicitly assumed that the advances in navigation and flight management system technology in the foreseeable future will be such that the defined 4D noise abatement trajectories can be tracked with great accuracy.

NUMERICAL EXAMPLES OF ENVIRONMENTALLY OPTIMIZED ARRIVAL TRAJECTORIES

As mentioned earlier, the NOISHHH tool has the capability to generate optimized noise abatement trajectories, either with or without including a time-of-arrival constraint in the problem formulation. In this subsection, we will show some typical results pertaining to environmentally-optimized approach trajectories, obtained without considering any time constraints (3D solution). This is then followed by some brief results related to a scenario, where a time-of-arrival constraint has been enforced (4D solution). In addition to results obtained for the composite performance index outlined in Subsection IV.B, we will also show some indicative results pertaining to an emissions-optimized arrival trajectory.

In the example scenario, a nighttime arrival from the Northern direction to runway 06 of Schiphol airport has been selected. As illustrated in Figure 11, this arrival route, which originates at the Spijkerboor approach fix, involves overflying several residential areas. In contrast to the Spijkerboor departure, the Spijkerboor arrival is not an officially published route. It is perhaps of interest to note that at present three CDA's to runway 06 are already in use. However, these CDA's are from different directions and do not originate at the Spijkerboor VOR.

The arrival trajectory to runway 06 is split up into five consecutive phases, each phase featuring a different flap setting. In the first phase, the aircraft flies a 3D flight path in a clean configuration, starting at a specified altitude and a given speed. The first phase terminates when the indicated airspeed reaches the flaps 1 extension speed (210 kts IAS). Moreover, it is assumed that at the end of the first phase the aircraft flight track is along the runway centerline extension. The second phase, flown with flap setting 1, relates to movement in the vertical plane only. It terminates when the speed for the next flap setting has been achieved, i.e., 190 kts IAS. The third phase is similar to the second phase except that it is now flown with flaps 5. Also the third phase terminates at a specified flap schedule speed, viz. 170 kts IAS. In the two subsequent phases, flight is assumed to take place along the glideslope. This implies that in these phases the flight path angle has been a priori set to -3°, thus further reducing the number of variables to be optimized. In the fourth phase flaps 15 is set and it is assumed that the undercarriage is in extended position. The fourth phase ends when the approach speed reaches 140 kts IAS. The fifth and final phase is flown with flaps 30 and ends when an altitude of 500 ft has been reached. No speed bleed-off is allowed in this terminal phase.

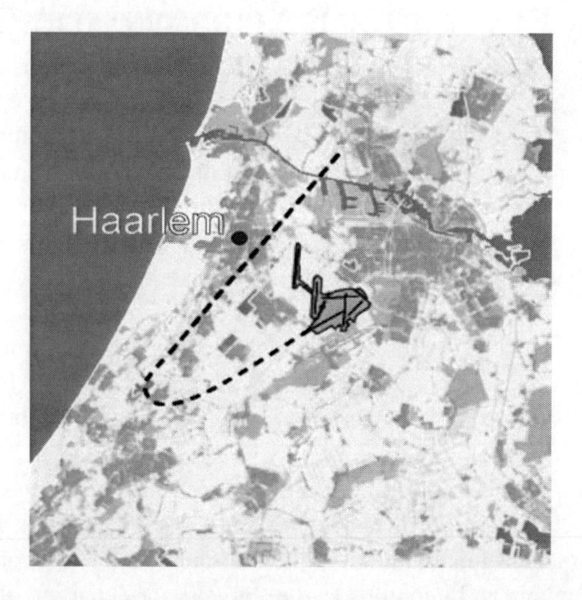

Figure 11. Runway 06 Arrival from the North at Schiphol airport.

The results are presented here in a coordinate frame fixed in space that has its origin at the grid point closest to the threshold of runway 06, with the x-axis pointing to the west (down-range) and the y-axis pointing to the south (cross-range). In each approach trajectory optimization calculation, all components of the initial state vector are specified. However, a complete parametric study involving the initial values of speed and altitude has been conducted in order to be able to identify the most favorable conditions. The specified initial heading corresponds to that of the existing Spijkerboor arrival procedure (Northern approach to runway 06).

Similar to the departure scenario considered in Subsection IV.E., a variety of constraints needs to be imposed in any given arrival scenario. The roll-angle limitations and speed constraints considered in the arrival scenario are similar to those adopted in the departure scenario. In order to obtain realistic results additional path constraints need to be included to inhibit (local) increase of speed (IAS) and altitude (h) during an arrival flight. Some care has to be taken in the multi-phase formulation to ensure a realistic behavior during the glideslope approach. Indeed, it is not to hard to conceive that without the introduction of appropriate operational constraints the optimization process is likely to result in a dead stick landing (unpowered). In order to avoid such unrealistic behavior, some altitude constraints have been imposed. These constraints include the requirement that the glideslope approach should be commenced at or above a specified minimum altitude. In the present calculations, the minimum altitude constraint has been (fairly arbitrarily) set at 2,500 ft AMSL. If so desired, this particular constraint can be relaxed to, e.g., 2,000 ft AMSL. A similar altitude constraint relates to the initiation of the final stage (flaps 30). In the present study, it is enforced that this final stage should be commenced at or above 1,200 ft AMSL.

In order to investigate the characteristic features of the optimal arrival trajectories, the principal parameters that have been varied are the weighting factor of the performance index, the initial speed, and the initial altitude. In this Subsection, the influence of the weighting factor K on the solution behavior will be examined for an assumed initial altitude of 7,000 ft, and an initial speed of 220 kts IAS.

By increasing the value of the weighting factor K in the composite performance index (see Subsection IV.B), the emphasis in the optimization process shifts from fuel-consumption to awakenings reduction. Figure 12 shows the optimal noise and fuel performance for a range of values of the weighting factor K. It can be observed that already a fairly large reduction in the number of awakenings can be obtained at the expense of only a slight increase in fuel-consumption, for moderate values of the weighting factor K. For values of K up to about 0.07 the increase in flight time remains limited (i.e., up to about 10 seconds). For values of K in excess of 0.07, flight path extension becomes more prevalent, with obvious consequences for elapsed flight time and fuel consumption. The specification of values of the weighting factor K in excess of 0.1 results in optimal trajectories that are partly located outside the defined noise grid area.

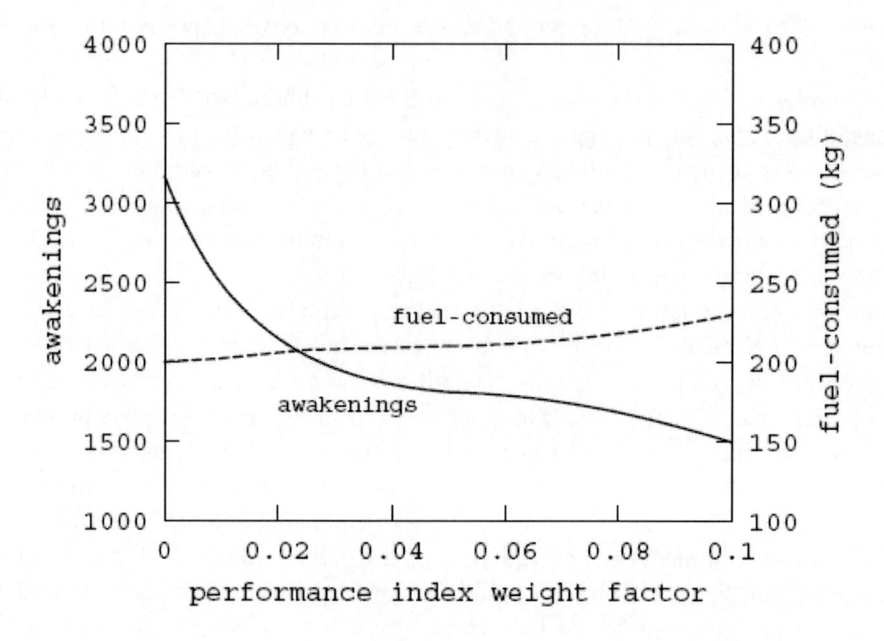

Figure 12. A comparison of optimal performances for various values of the weighting factor K.

In Figure 13(a) the optimal trajectory solution pertaining to the minimum-fuel (K= 0) scenario is shown. Each of the four graphs included in Figure 13(a) shows the movement of the arrival aircraft in three-dimensional space, along with the projection of the flight-path onto the horizontal plane (ground tracks) as well as contour plots within the noise grid area related to one of four selected metrics, viz., outdoor Sound Exposure Level (SEL), the percentage of people expected to awake, the population density distribution, and the number of people expected to awake as a result of the flyover. Figure 13(b) presents similar graphs for a noise-optimized trajectory, obtained for K= 0.1

A comparison of the optimized trajectories presented in Figures 13(a) and 13(b) reveals that the largest impact of the performance index weighting is found in the shape of the downwind arrival route. Indeed, in contrast to the minimum-fuel route which directly overflies the large city of Haarlem, the noise-optimized solution essentially flies around this community. Moreover, the thrust setting is set to flight idle when the aircraft flies close to the outskirts of Haarlem. This early reduction in thrust in the minimum-noise solution obviously must be compensated in the sense that additional power needs to be added at a later stage. This explains why there is an additional area enclosed by the 2% relative awakening contour line in the noise-optimized solution. Since most of the 2% relative awakening contour is located in uninhabited areas, the impact on the absolute number of awakenings is rather modest. The overall improvements in noise impact that can be obtained are fairly impressive. Indeed, the number of people that awake due to noise impact reduces from 3,166 for the fuel-optimal trajectory to 1,495 for the noise-optimal trajectory. In relative terms this implies a reduction of more than 50%. With respect to fuel consumption, the differences between the minimum-fuel and the minimum-noise trajectories are rather modest. The noise-optimal trajectory requires only about 30 kg (or about 15%) more fuel. The additional path length of the horizontal route in the minimum-noise solution obviously has its impact on the elapsed flight time. The fairly extensive detour taken in the minimum-noise solution leads to an additional flight time of about 50 seconds (or about 10 %).

When the results for the arrival flights are compared to the corresponding departure results (presented in Subsection IV.E), it needs to be realized that in an arrival scenario relative more people are exposed to (low-intensity) noise in comparison to a departure scenario. This is also reflected by the fact that the noise grid area size adopted in the arrival scenario is more than twice the size of the area considered in the departure scenario.

To illustrate how the improvements in noise performance are brought about, some more detailed trajectory optimization results are shown in Figure 14. First of all, due to the low initial energy situation, the downwind arrival leg is not completely flown at flight idle. In the minimum-fuel solution, some thrust is added in the initial phase, where the prevailing (high-altitude) flight conditions are more fuel-efficient. In the noise-optimized solution, the initial thrust setting is reduced much earlier in order to alleviate the nuisance in the residential communities that are overflown. This early thrust reduction is accompanied by a higher descent rate. Once the residential areas have been crossed, thrust is slightly increased and the flight path becomes shallower again. In the glideslope phase the solution behavior is not dramatically different for the two cases.

An interesting feature present in both solutions is the horizontal flight path segment that occurs just prior to glideslope interception. The existing procedure also features a horizontal segment, but in contrast to the optimized trajectories, this horizontal segment is not flown at idle thrust. The idle-thrust horizontal segments in the optimized solutions enable a very quick deceleration, which helps to limit the overall path length. Note that during the horizontal segment the flaps are gradually extended. The flap setting transitions (and thus the different flight phases) can be seen in the time-history of IAS.

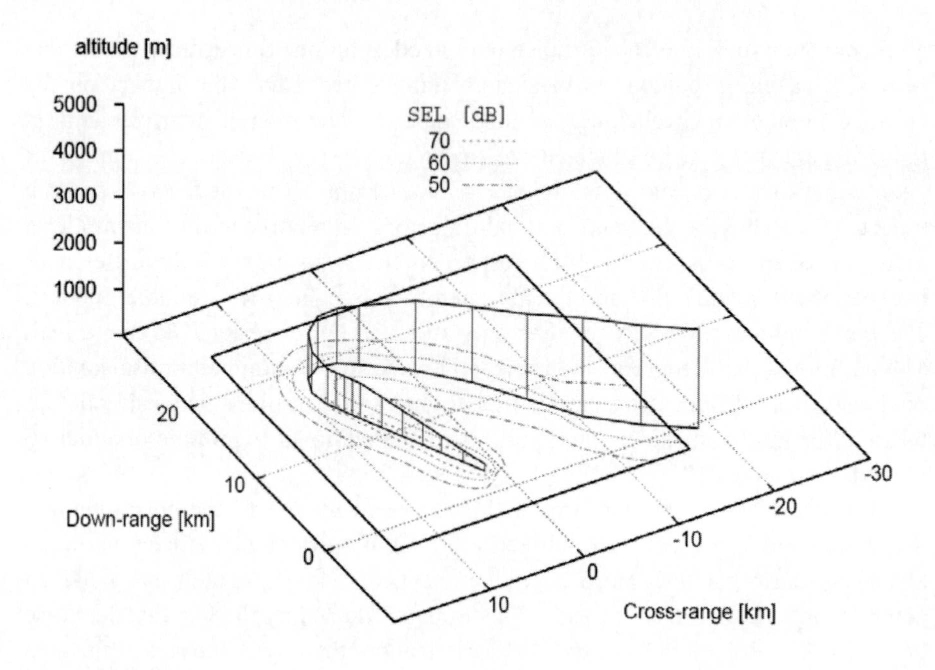

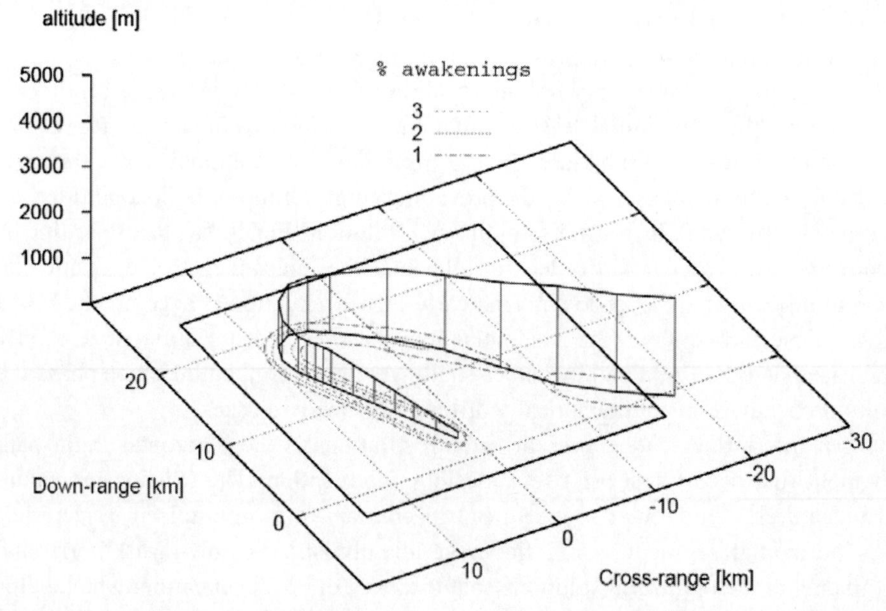

Figure 13a. (Continued)

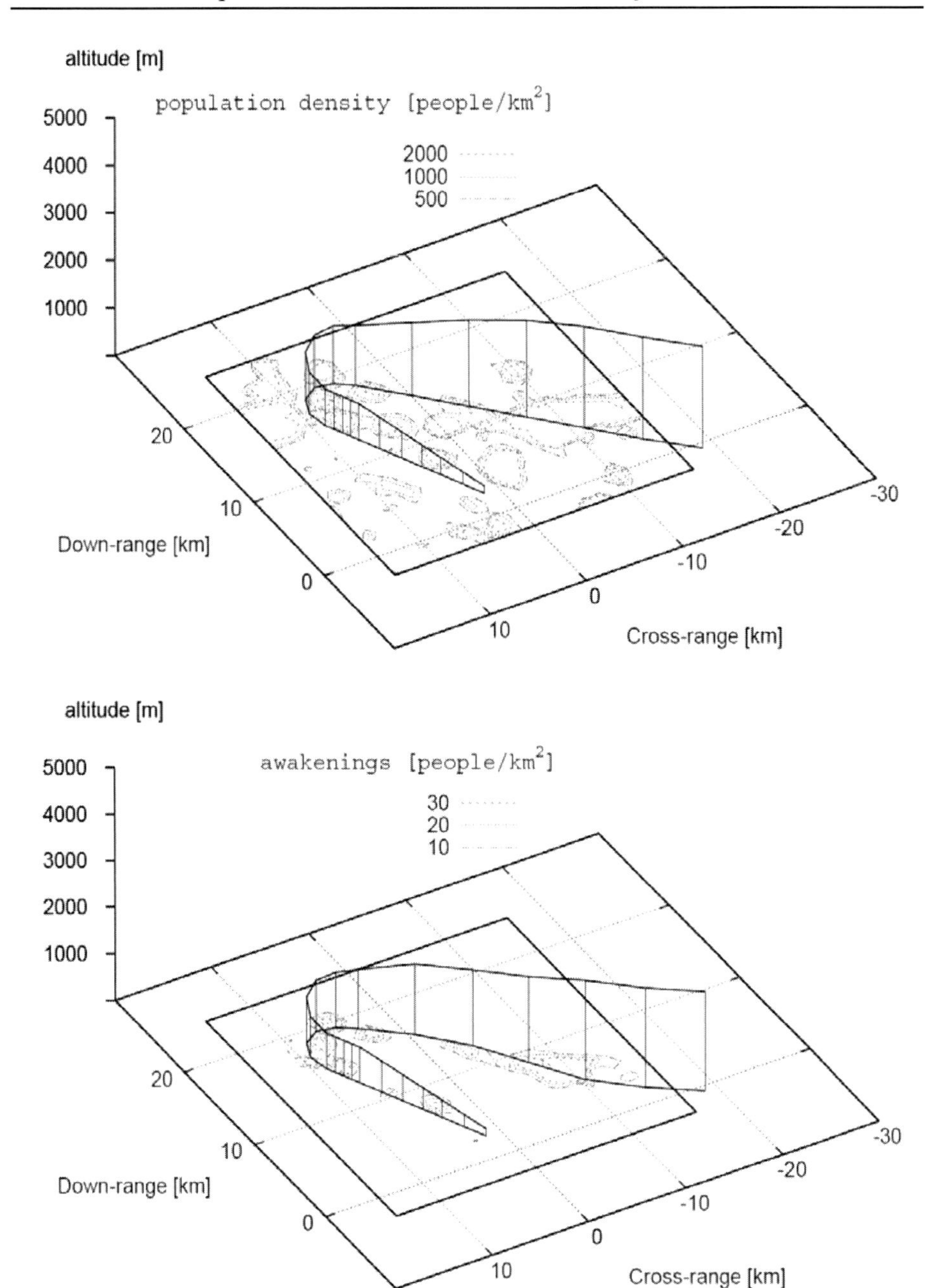

Figure 13a. Minimum-fuel arrival trajectories.

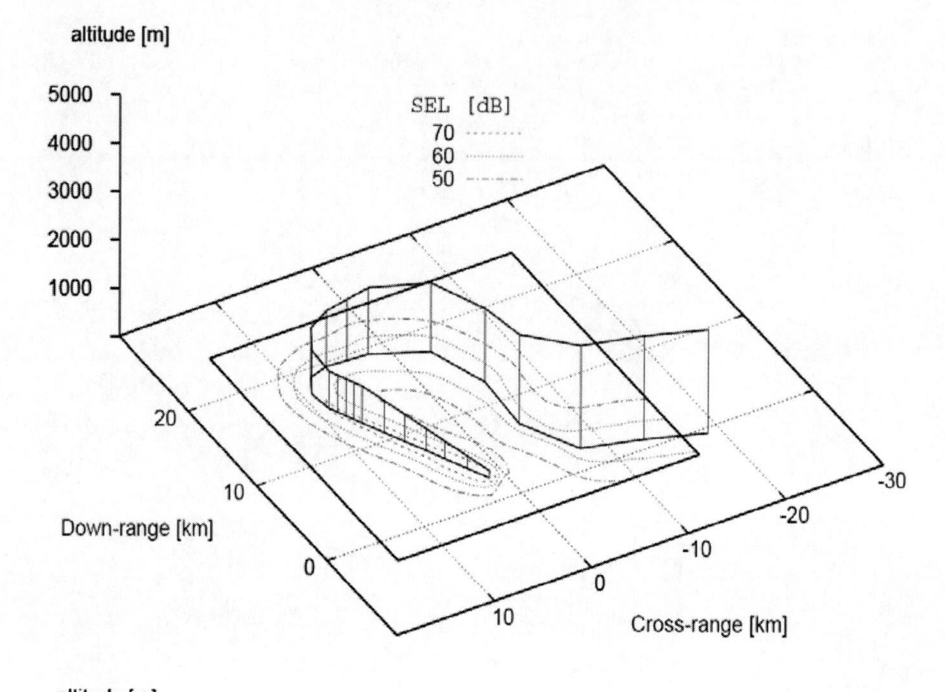

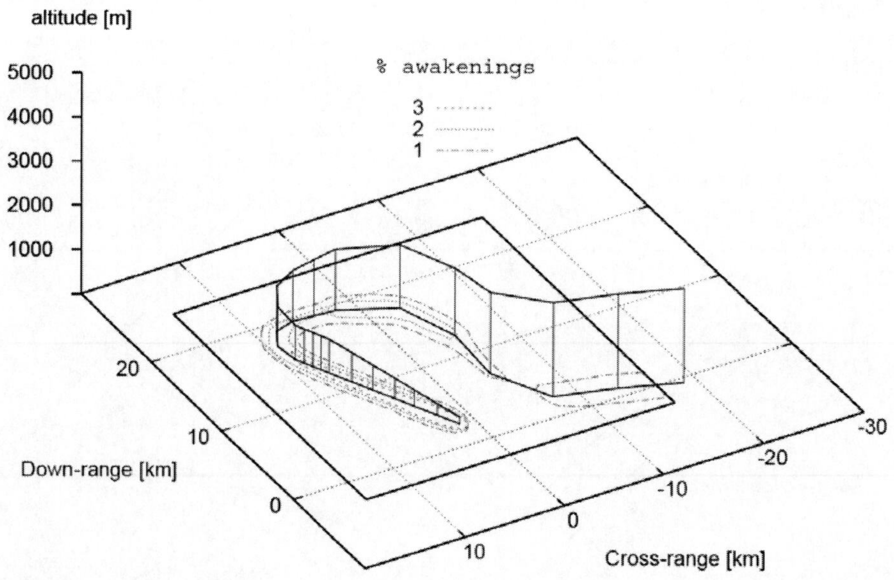

Figure 13b. (Continued)

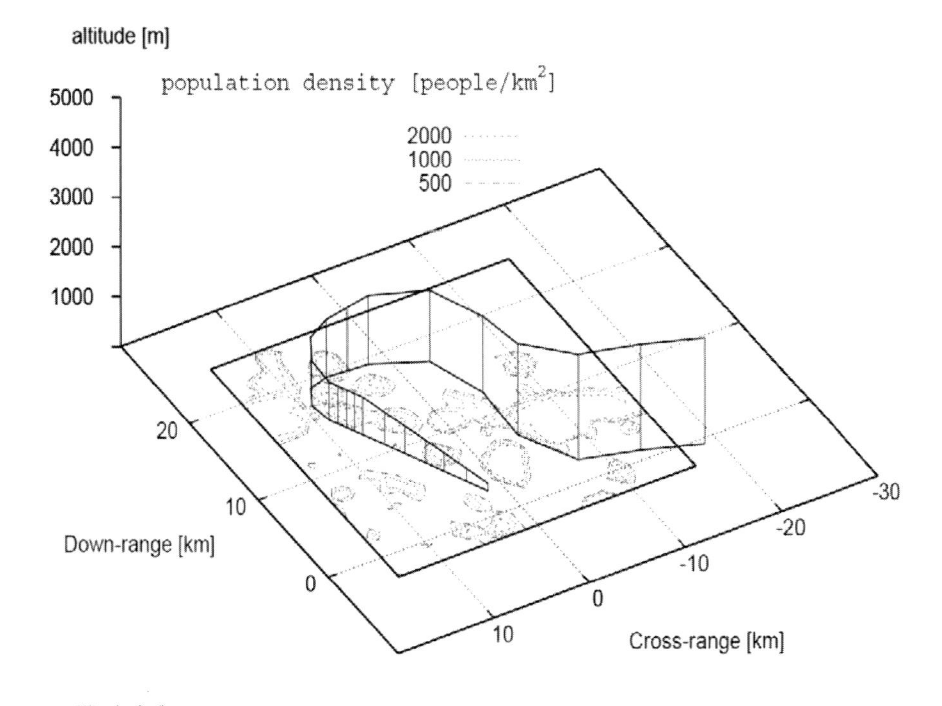

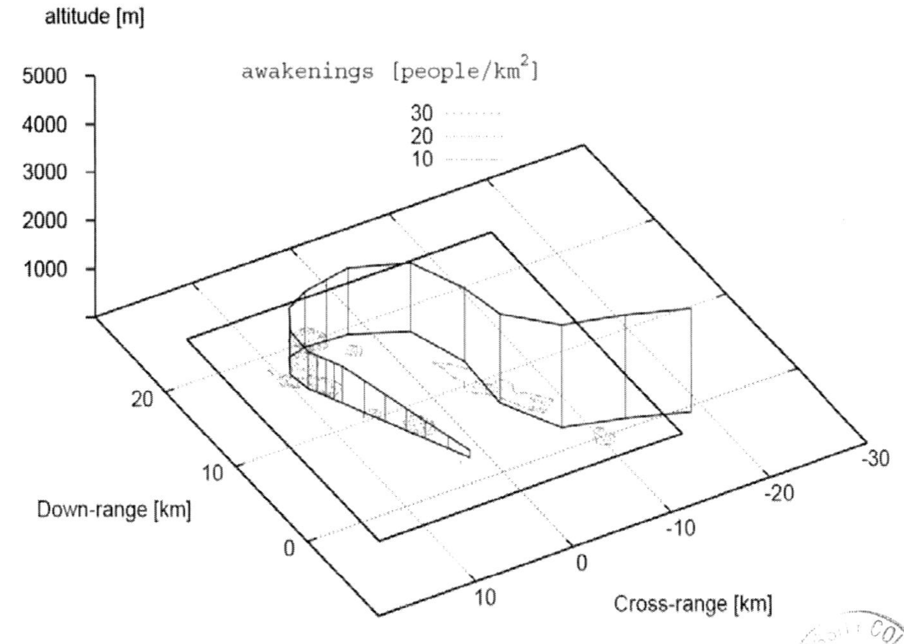

Figure 13b. Noise-optimized arrival trajectories.

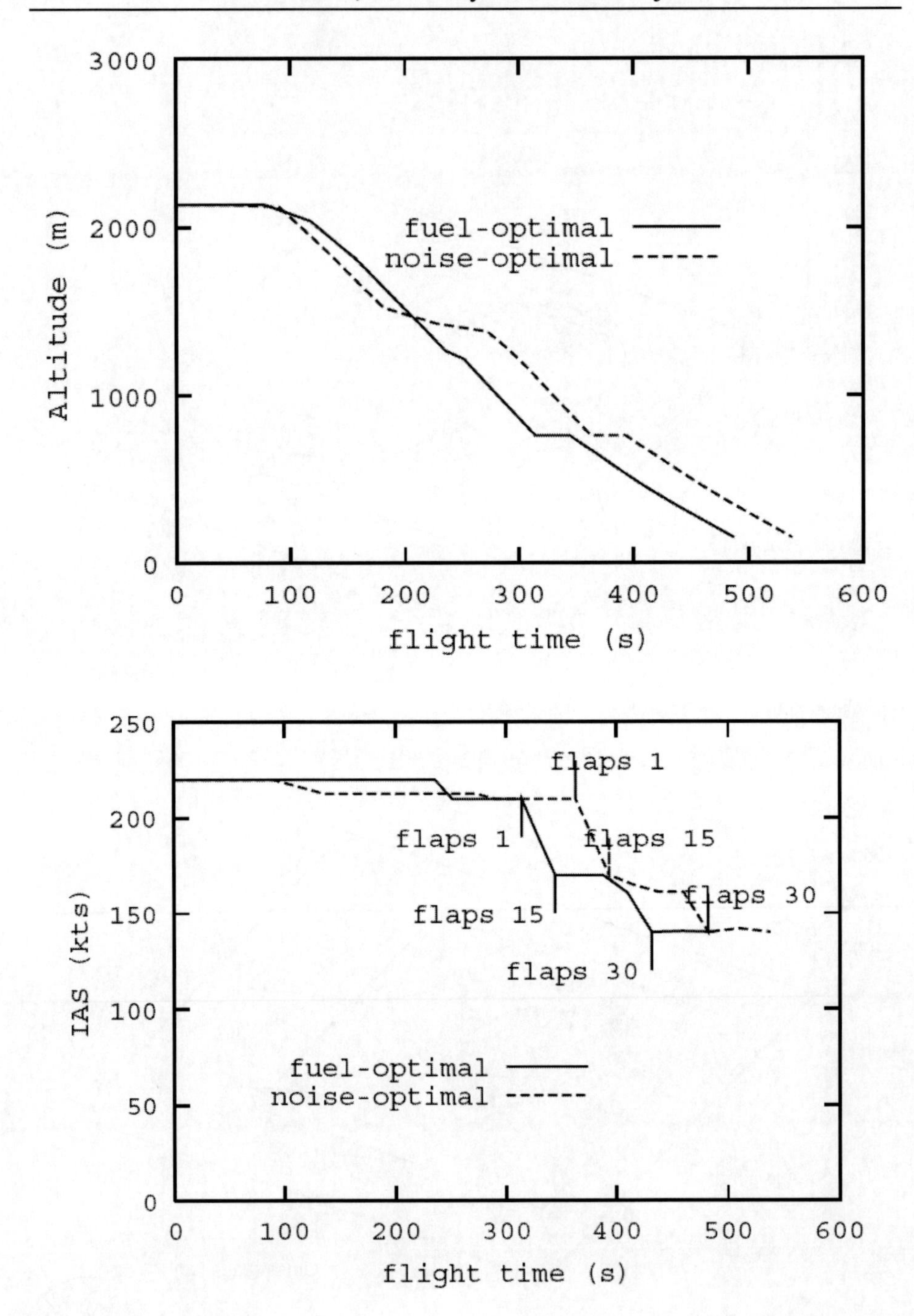

Figure 14. (Continued)

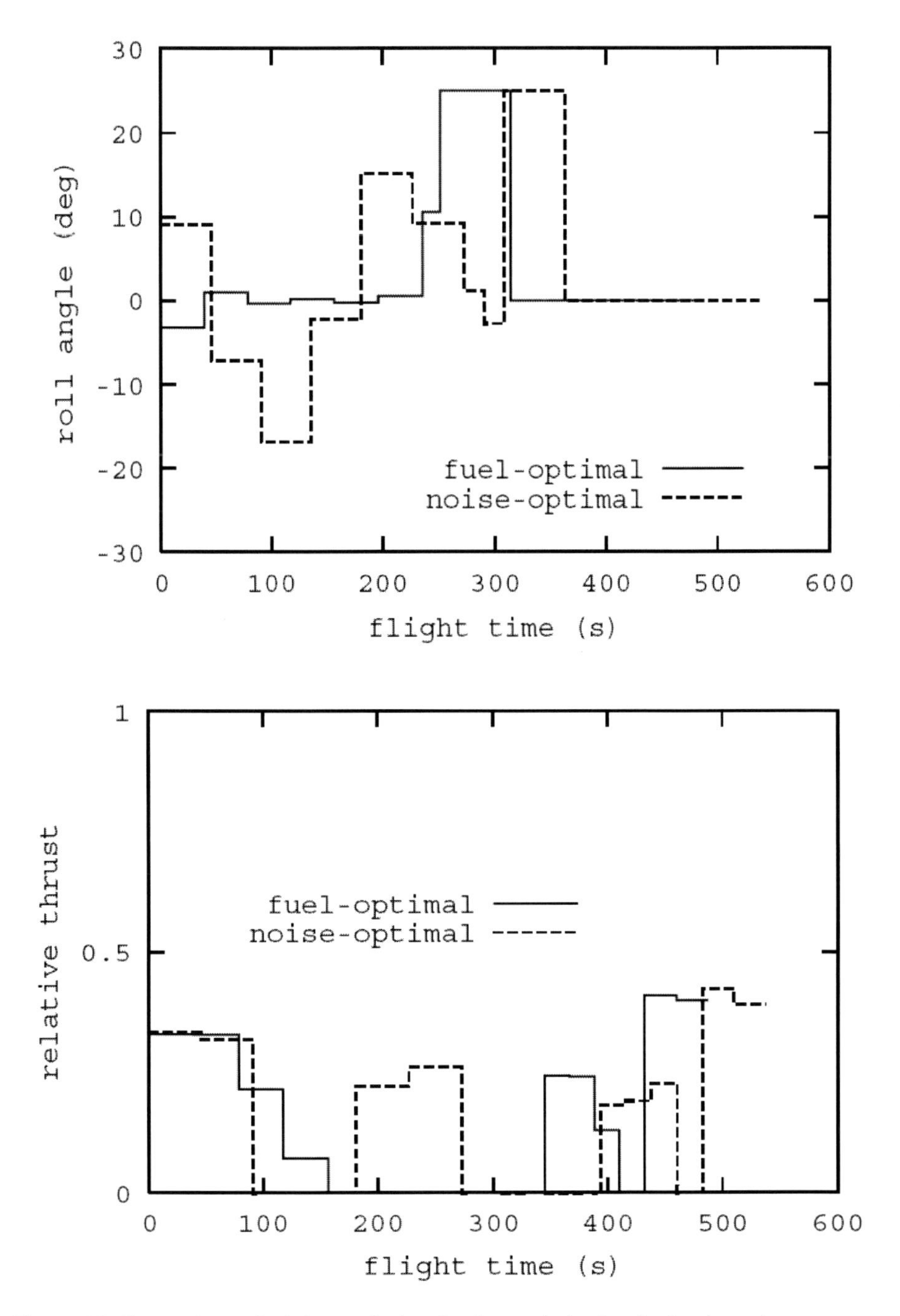

Figure 14. Comparison of minimum-fuel and noise-optimized arrival trajectories.

In the next example, we examine the influence of higher initial altitudes prior to the start of the approach/arrival procedure. The trajectories in this example have been computed for a moderate value of the performance index weighting factor, i.e. $K = 0.01$ Figure 15 summarizes the benefits of a higher initial altitude in relation to fuel-consumed and expected awakenings. Increasing the initial altitude to a value in excess of about 11,000 ft is no longer beneficial with respect to fuel consumption, in the sense that flight path stretching will be required to absorb excess energy. Moreover, since path stretching will essentially take place over the North Sea, the noise benefits will level off with increasing initial altitude.

To be able to investigate the influence of initial speed, the scenario pertaining to Figure 15 has also been considered without the imposition of the 220 kts IAS limitation, allowing the initial speed to be increased to 250 kts IAS. The arrival trajectory optimization is not only expected to benefit from an increase in potential energy, but from an increase in kinetic energy as well. The optimal solution for the scenario with increased initial speed indeed results in 357 fewer awakenings and 14 kg less fuel-consumed, in comparison to the scenario featuring the lower 220 kts initial speed. The results obtained in this example appear to indicate that an optimization of desired altitude and speed at the initial approach fix for each specific situation is clearly warranted.

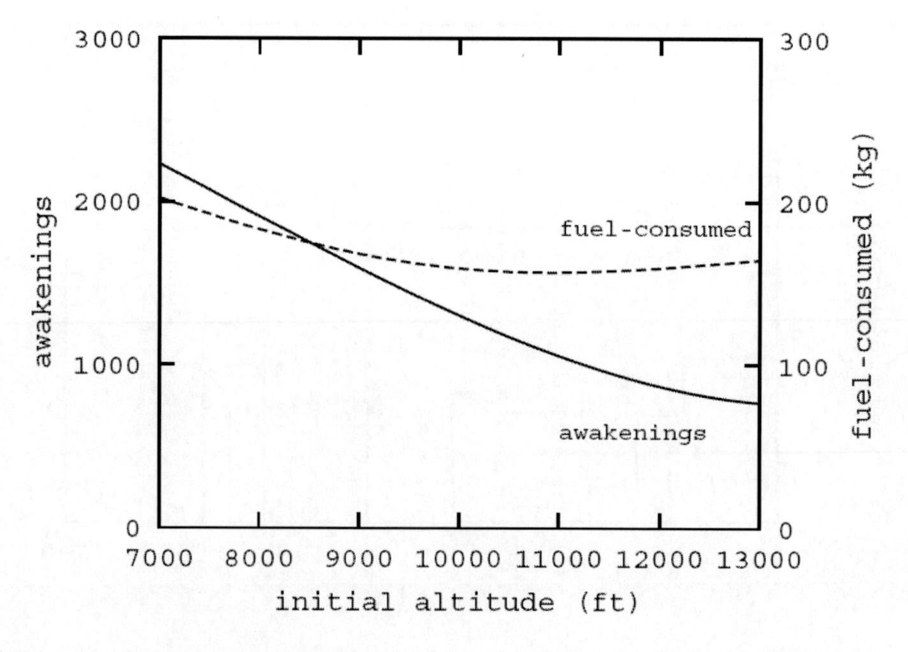

Figure 15. Comparison of optimal performances for various initial altitudes (K= 0.01).

To demonstrate the feasibility of 4D trajectory optimization, the scenario corresponding to the noise-optimized solution shown in Figure 13(b) (with $K = 0.1$) has been re-assessed, with a flight time constraint added. Figure 16 shows the expected number of awakenings and fuel consumption performance for a range of values of the prescribed transit time. Inspection of Figure 16 clearly shows that a short transit time is unfavorable from both a noise and fuel (emissions) perspective. It is noted that the solution with the lowest fuel consumption that can be extracted from Figure 16 differs from the true minimum fuel solution ($K = 0$), as presented in Figure 13(a).

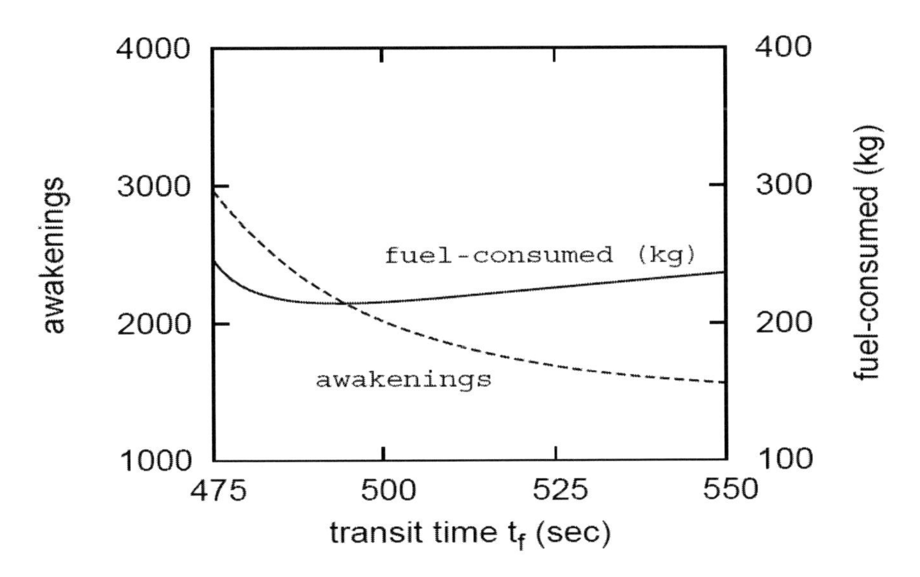

Figure 16. A performance comparison of noise-optimized arrival trajectories for fixed flight times.

The final numerical example concerns the optimization of an arrival trajectory with respect to emissions. In contrast to the previous examples, the primary performance index considered in this example is not related to noise, but rather to the marginal external costs of emissions. The external costs of air pollution are based on an economic valuation of the effects of primary pollutants on the health of human beings as well as on crop production. The environmental damage costs considered here are based on pollutants that are emitted below 3,000 ft AGL. For the part of the arrival trajectory above 3,000 ft AGL, fuel burn is used as the optimization criterion in this particular example. The external costs of emissions are based on the valuation data for emissions in rural area in The Netherlands [Holland, 2002] and are summarized in Table 1:

Table 1. Marginal external costs in rural areas in The Netherlands, year 2000 prices [Holland, 2002]

	SO_x	NO_x	HC
Cost (€/tonne)	7,000	4,000	2,400

The results for the arrival trajectory optimization based on the overall the emissions damage cost criterion are summarized in Table 2. It can be observed that the differences between the minimum-fuel solution and the emissions-optimized solution are only marginal. The emissions-optimized solution only features a slight reduction in the marginal external cost of only about 1% relative to the minimum-fuel solution. The cost improvement is achieved through a reduction in the emission level of NO_X; for the remaining pollutants the emitted masses are even slightly increased. The primary reason for the fact that the differences in emission levels between the two solutions are so small relates to the fact that most of the pollutants are emitted during the glideslope phase, which is highly constrained and leaves little room for thrust variations.

Table 2. Comparison of performances of minimum-fuel and emissions-optimized arrival trajectories

Criterion	Fuel used (kg)	Awakenings	Transit time (s)	NO_x (g)	HC (g)	SO_2 (g)
fuel	202	3,166	490	2,402	7.86	169.9
emissions	203	3,170	494	2,360	7.92	170.5

CURRENT AND FUTURE WORK

Although the noise-optimized trajectories produced by the current version of NOISHHH reveal a huge potential for noise impact reduction, they do not readily lend themselves for direct application to terminal area routing, at least, when relying on existing technology and procedures. The presently obtained noise-optimized approach and departure trajectories are rather complex in nature, in the sense that they exhibit fairly extensive lateral maneuvering, multiple speed changes, as well as significant variations in the descent rate or climb rate throughout the trajectory. For this reason NOISHHH is currently being modified in such a way that it is able to generate optimized trajectories that are compatible with the RNAV procedures currently in existence [ICAO, 1999]. RNAV can be loosely defined as a method of navigation that permits aircraft to fly any desired (pre-programmed) flight profile (within technical and operational constraints), without the need to fly directly toward or away from a ground-based navigation aid (beacon). The first generation of RNAV systems only permitted accurate navigation within a horizontal plane (2D RNAV). Modern RNAV systems, which are usually integrated within the FMS, also provide vertical plane navigation (3D RNAV) or even time-based navigation (4D RNAV).

Every RNAV arrival or departure horizontal flight path is built from one or more segments. Each RNAV route segment contains two basic elements, (i) a waypoint, which is a specific location defined by latitude and longitude coordinates, and (ii) a leg type, which defines the path before, after or between waypoints. Although a large number of leg types is currently available, for the design of flight routes in NOISHHH only two basic leg types are used. The first one is a straight leg between two waypoints, called a "Track to a Fix" (TF) leg. The other leg, called the "Radius to Fix" (RF) arc, defines a constant radius turn between two waypoints, with a fixed radius around a given center point.

The vertical flight path is constructed by connecting the speed and altitude constrained waypoints along the horizontal flight path defined by the RNAV route. The vertical navigation mode (VNAV) of the FMS allows building a vertical path between waypoints in several ways; one of the possible vertical navigation modes guides the aircraft from waypoint to waypoint with a constant flight path angle. This particular geometric approach of constructing a point-to-point vertical path has been presently adopted in the RNAV-modified version of NOISHHH, primarily due to its generic nature. In addition to speed and altitude constraints, also time-constraints can be imposed at the waypoints, so that essentially a 4D trajectory solution is obtained.

Figure 17 shows the impact of including a set of RNAV requirements in the optimization formulation on the ground track behavior, for a Northern approach to runway 06 at Schiphol airport in The Netherlands. This particular solution has been obtained for a value of the weighting parameter $K = 0.1$ in the composite performance index (see Subsection IV.B)

A close inspection of Figure 17 reveals that the RNAV solution only features two constant radius turns and therefore does not avoid the populated areas as well as the original solution. As a result, the RNAV solution leads to an increase in expected awakenings of about 6% (93 people) relative to the original solution.

Due to the fact that the length of the overall track is shortened, the fuel consumption is improved by about 1% (2 kg) in the RNAV solution.

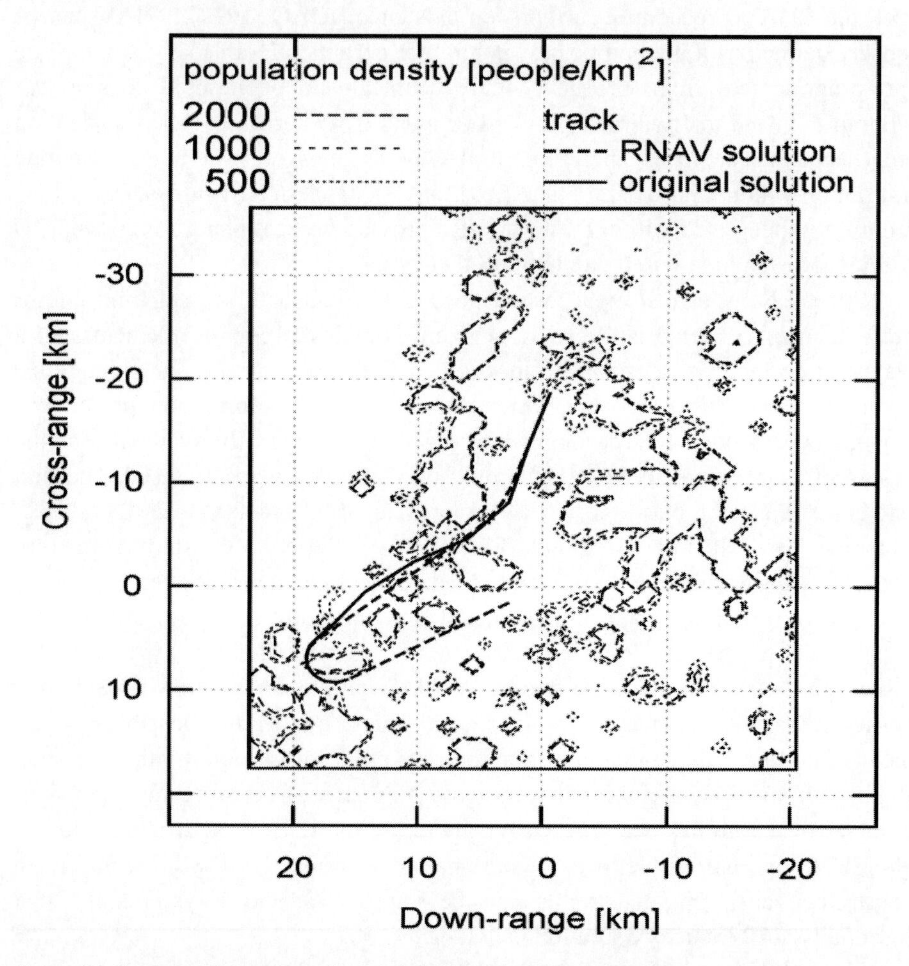

Figure 17. A comparison of a noise-optimized RNAV arrival trajectory with the corresponding original NOISHHH solution.

A second modification of NOISHHH on which we currently work, relates to an extension of NOISHHH from a single event tool to a dual event tool, with the aim to assess the effect of resolution decisions related to in-trail separation conflicts on noise impact. In essence, the dual event tool assesses the combined single event noise impacts of two in-trail arrival trajectories that feature a separation conflict. The trajectories can be optimized with respect to a defined environmental criterion, either with or without resolving the separation conflict. A comparison of the results enables not only to quantify the noise impact of a conflict resolution, but also to identify the essential characteristics of the optimal resolution strategy. Initial results [Visser, 2006] appear to indicate that conflicts are primarily resolved by altering the flight characteristics in the vertical plane (viz., altitude, speed and thrust profiles) rather than through flight path modifications in a lateral sense. In the numerical trials conducted thus far, two different aircraft types were used (Fokker F100 and Boeing 737, respectively), each of which contributes about equally to the defined environmental performance index. As a result, both aircraft types contribute to the conflict resolution in an equitable fashion. A first next step in the research effort will involve the assessment of optimal conflict resolutions for a wide variety of aircraft types, such as to assess to what extent differences in aircraft environmental performance affect the optimal conflict resolution behavior.

To date our research has consistently relied on the assumption of standard atmospheric conditions and no wind. In the near future we intend to direct research efforts towards the assessment and modellling of the effect of weather on aircraft noise impact and to explore how the obtained insights might affect changes in flight procedures.

Chapter 4

ENVIRONMENTAL PLANNING AT THE STRATEGIC LEVEL

LAND USE PLANNING

Environmental planning at the strategic level is a long term planning activity that is carried out to achieve certain environmentally related goals in the future. With respect to airports, land use planning, also referred to as spatial or urban planning is probably the most effective instrument in the strategic range. When performed properly, it ensures that the land use near the airport remains compatible with the airport's operation, thereby preventing future conflicts. Typically, this is government responsibility and achieved through the issuing of development restrictions.

Noise load is a primary factor in determining whether or not a particular location is suitable for construction. Important is that planners should not only look at the current noise load, but also need to keep in mind how noise load may develop over time. There is no universal or absolute noise limit. Regulations not only differ among countries, sometimes different regulations are used within the same country depending on the status or function of the airport. The nature of construction is important as well. Lower noise limit guidelines may apply for noise sensitive buildings like schools, hospitals and libraries when compared to offices, shops and plants [FAA, 1983]. Often existing construction can be maintained at noise loads where new construction for the same destination would not be allowed, or at least not without additional requirements on the acoustic insulation of the building.

Third party risk can also be a concern when performing spatial planning. Risk of aircraft impact may be deemed too high for certain areas, especially those near the runway ends. Just like with noise, areas that are found to be unsuitable for a certain type of development are not automatically destined to lay fallow. For example, in the Netherlands, areas with a risk too high to be used for housing can still be used for commercial purposes as long as the expected people density is low. This means that it can still be used for most agriculture activities or maybe a long-term parking facility for the airport as well.

Even though land use planning is used, incompatible land use may still occur. In the case of noise compatibility, situations like that arise when for example the noise load is higher than anticipated, or when limits are lowered. In such a situation, several options exist. First of all, the situation may simply be tolerated, which usually works fine as long as none of the stakeholders opposes. Second, reduction of the noise load is an option, but this may not be possible without severe airport access restrictions, with resulting damage. Next, noise insulation, or sound proofing can be an efficient solution, especially when limits take insulation into consideration or are defined as an indoor maximum. Finally, the drastic solution of (involuntary) property acquisition remains. In this case, it is usually the airport that compensates the owner by buying the property and demolishing it.

AIRPORT STRATEGIC PLANNING

Apart from land use planning - primarily a government responsibility - environmental issues often need to be taken into consideration during airport strategic planning projects. However, airport management does not have direct control over the environmental impact. Apart from access restrictions, demand shaping is one of the instruments used. Several European major airports use noise related landing fees or penalties. This can discourage operators only owning the noisiest aircraft types from flying into such an airport, and stimulate operators that have a choice in using a less noisy type out of their fleet. On a longer term, it may also have some impact on fleet replacement decisions, because noise performance becomes a factor in the operating economics of the aircraft.

Demand shaping can also be more subtle, as is shown using the following example: Consider an airport that is facing problems with respect to their capacity. Especially the (physical) capacity of the terminal buildings poses a problem, but the environmental capacity with respect to noise will become problematic as well with further growth. Now assuming that the terminal capacity problem can be solved through an expansion project, the question arises if it is interesting to invest in a terminal building for a particular segment of the market considering the tight environmental capacity. For example, if the current freight operations use a disproportional part of the noise budget compared to its volume or revenues, investing in passenger facilities only might be the wiser option, because it is expected to allow for more growth within the remaining environmental capacity.

HARMOS

Environmental impact is only one of the issues airport planners need to consider in major strategic expansion projects. To name a few other focus areas, financing, land acquisition, operational efficiency, safety and overall timing need to be considered. Because of the different disciplines, multiple experts provide the required information with respect to their fields of expertise. This means that planners need to combine the information from different sources and find solutions while meeting often contradicting requirements. To aid them in identifying efficient solutions, a decision support system (DSS) for airport strategic planning is currently under development at TU Delft. This system is called HARMOS [Wijnen, 2006]. It is able to work with a whole range of modelling tools, currently used by experts from the different domains. Several optimization engines use the results generated by the modelling tools to come up with promising solutions. Apart from the efficiency gains that can be expected from the optimization approach, another advantage is the use of a single dataset. Because all the input for the different analysis models is generated from a single data model, consistency can be guaranteed.

One of the optimization models available to HARMOS focuses on the environmental effects of runway usage, or otherwise states, the noise load and third party risk resulting from a certain distribution of the flights over the different runways of an airport. The tool, dubbed Strategic Noise Allocation Program (SNAP), allows the user to assess and optimize the runway usage on a strategic level [Hebly, 2005]. For the remainder of this section, we will focus on this particular interactive optimization tool. It should be noted that although the tool was originally developed for HARMOS, its concept is also a candidate for the strategic modules of the integrated environmental management tool.

Strategic Noise Allocation Program

Larger airports with several runways usually have multiple modes to operate in. A particular mode, called a runway configuration, describes for each of the runways whether they are open, used for arrivals, departures or both simultaneously and in which direction they are used. Different configurations result in different traffic patterns around the airport, in turn resulting in a different noise load and third party risk distribution. Although total noise load and total third party risk is not influenced by runway configuration selection, the geographical allocation of these environmental effects certainly is. Especially if there are both sparse and very dense populated areas near the airport, the resulting risk and experienced noise load can differ substantially among the different operating modes.

However, when selecting a runway configuration, environmental effects are not of primary importance. Typically, weather conditions prevent most of the configurations from being used because of safety considerations. Some of the remaining configurations might not offer the appropriate arrival or departure capacity for the expected traffic levels, preventing them from being selected as well. It is very well possible that at this stage, only one possible configuration remains. However, if multiple configurations are eligible, this is the point where environmental considerations may be taken into consideration.

The problem described here is modelled in SNAP. It can optimize the allocation of flights to runways on an annual basis with respect to noise, risk and average delay, while respecting constraints related to operational procedures, required capacity and weather conditions. The input parameters are controlled through HARMOS. Using scenarios, it is possible to analyze the effects of changes in the input. Typical scenarios involve an altered runway system, changes in traffic type and levels or even a changing climate.

Optimization Model

When setting up a new optimization, SNAP starts by collecting all required data. This includes a detailed traffic schedule, available runway configurations, statistical weather data and population data. Based on the traffic schedule, the noise and third party risk contributions for all aircraft types are calculated for each runway and routing combination. For noise, this is done using INM, calculating the sound exposure level for each possible movement (see subsection II.C). For third party risk, the NATS risk model is used, calculating the individual risk contribution for each conceivable flight (see subsection II.E). These results are stored for later use.

Next, the traffic schedule is scanned on an hourly basis to look for similar traffic patterns throughout the year. A similar process is started on the weather data, identifying all weather situations that allow for the same selection of runway combinations. For this version of SNAP, only tail- and crosswind conditions are considered, as the functionality for visibility conditions has not yet been implemented. When statistically combining results from the traffic and the wind analysis, the year is divided into a number of periods, where each period has distinct traffic demand and range of available runway configurations. Each of the periods has a certain duration, based on its statistical occurrence. Of course, the duration of all periods together should yield a full year.

The actual optimization model is a Mixed Integer Programming (MIP) model. For each of the periods, the solver is forced to select from a number of options. Each option represent a certain runway configuration and if applicable, also a distribution of the traffic over the different active runways of the configuration. Since only feasible configurations are offered to the solver and the solver is always able to select one of the options offered because there are no constraints preventing that, the solution will always be feasible, both mathematical as well as operational.

The selection of an option for a particular period translates into a distribution of flights over the runways. This automatically determines the contribution for noise, risk and delay for that period. In the optimization model, these factors are modelled as three separate cost or penalty functions. Because there are three different cost functions, the problem is multi-objective and will probably not have a unique solution. Unless the three objectives will never conflict, the final solution will be dependent on a trade-off between the three criteria. In order to reach a final solution, an interactive procedure has been developed for SNAP. A schematic overview of the total model is given in Figure 18.

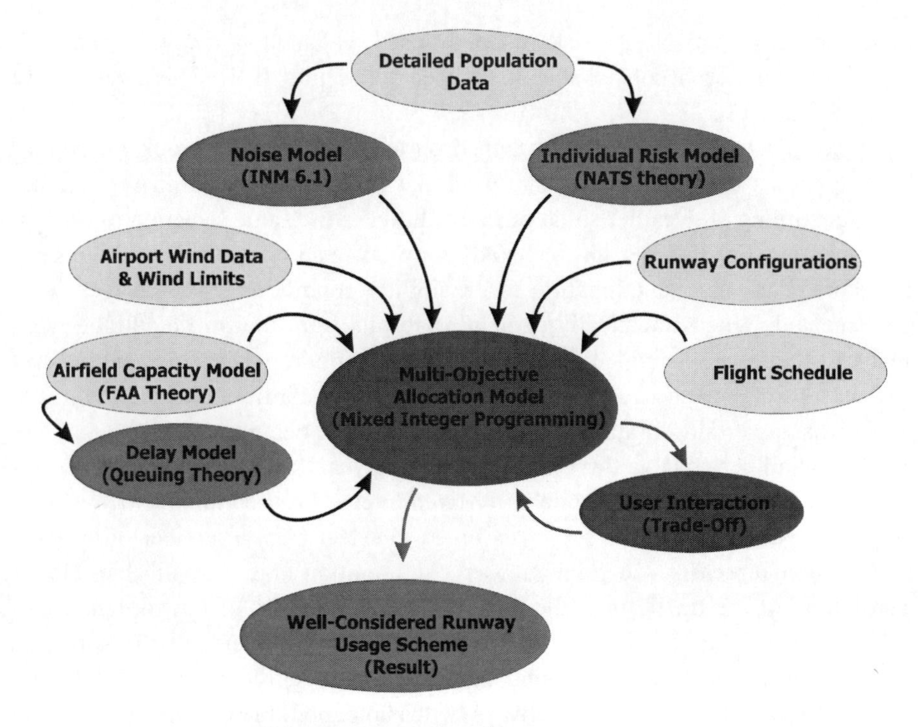

Figure 18. Design of the optimization module.

INTERACTIVE OPTIMIZATION PROCEDURE

The interactive procedure is based on a weighted sum method, where the final objective is a linear combination of the three cost functions [Deb, 2001]. However, compared to the weighted sum method, the user does not need to specify the weights beforehand, which requires a-priori knowledge of the optimization. Especially for non-experienced users, this can be a problem, making it hard to reach a satisfactory result.

Instead, the interactive procedure first determines the weights automatically such that all three objectives become equally important at the point where they reach their absolute minimum. These values are obtained by solving the problem three times, each time with only one of the objectives present in the goal function. Next, the decision maker is shown the ranges for the three objectives, together with the three initial solutions. Starting from one of these initial solutions, a new solution can be generated that will minimize for all three objectives simultaneously.

In generating a new solution, the decision maker can specify upper bound on all three objectives, based on information obtained from previous solutions. In practice, it is easiest to relax achieved values for one objective, if the other two remaining values should be improved. This is illustrated in Figure 19. It shows the user interface used during the interactive optimization procedure. The upper part presents the minimum delay solution, one of available initial solutions. The lower part of Figure 19 shows the solution that has been obtained from the minimum delay solution after a single iteration. The previously obtained value for average delay has been relaxed by a few seconds, allowing for the noise and risk values to be improved significantly.

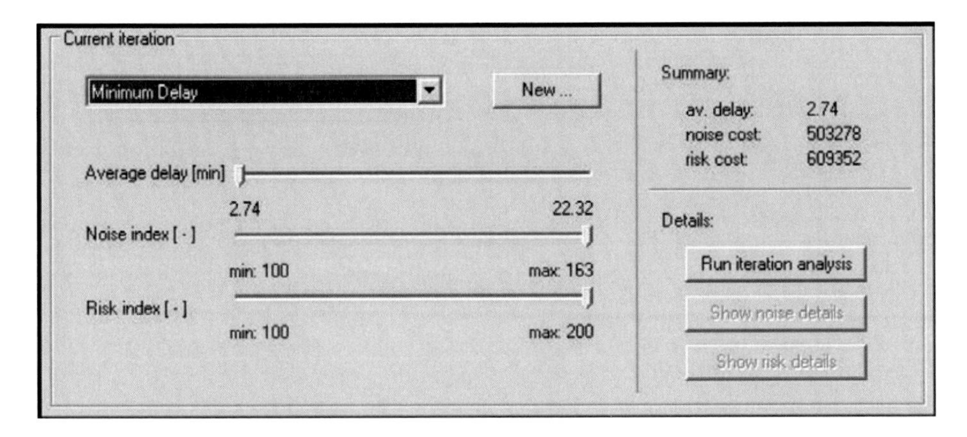

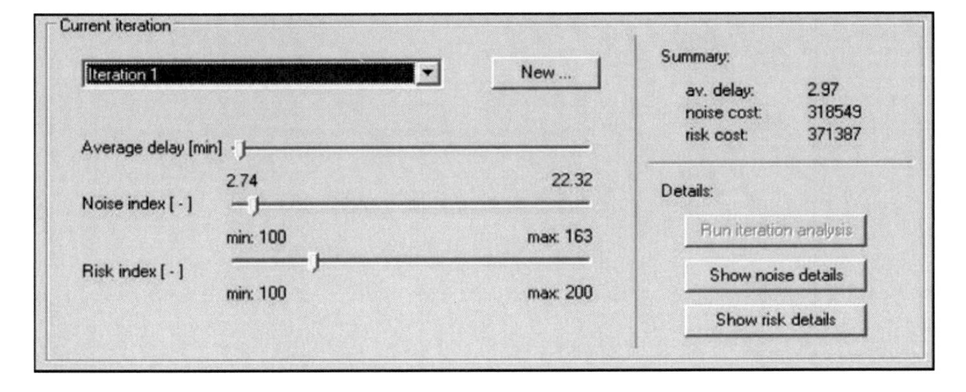

Figure 19. The user interface used for the interactive optimization procedure.

EXAMPLE RESULTS

To show the effects of runway allocation optimization, results generated by SNAP are compared to a reference scenario, concerning Amsterdam Airport Schiphol. This reference scenario is also calculated by HARMOS, based on operational data. Both the reference and the optimized results use the same amount of traffic and calculation methods. The only difference is that the reference case uses runway usage percentages as projected by the airport authority, where SNAP optimizes the runway usage according the preferences of the user.

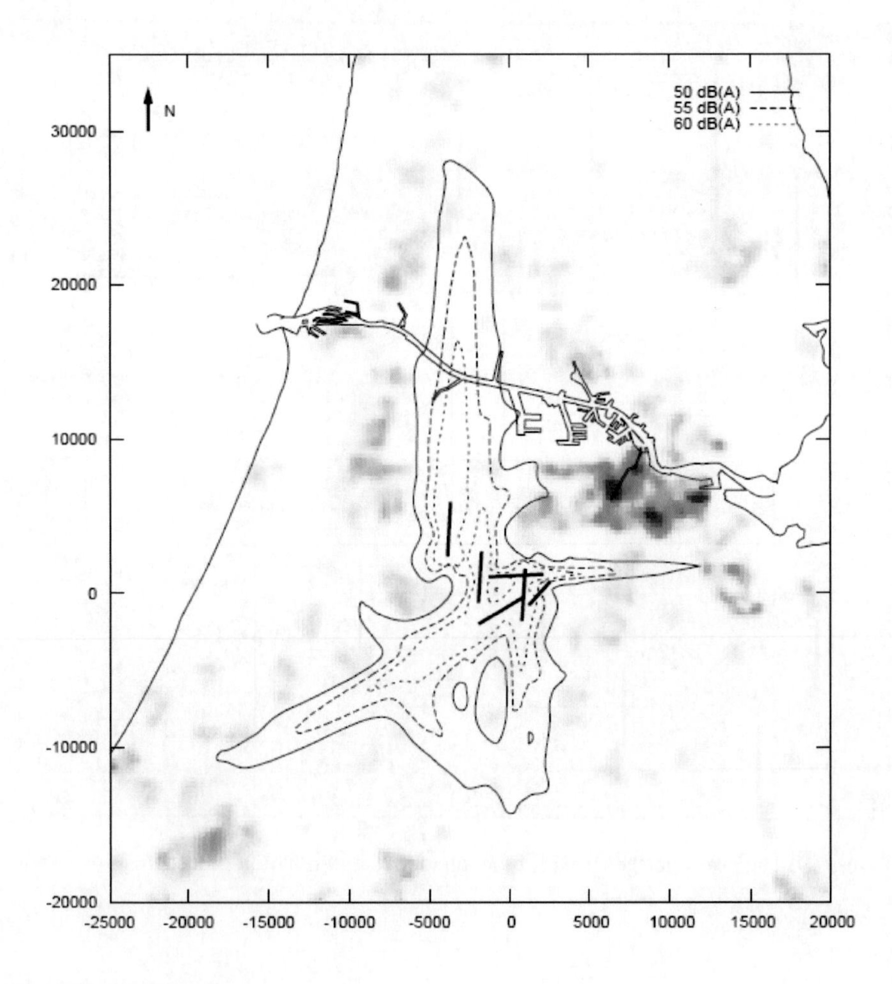

Figure 20. (Continued)

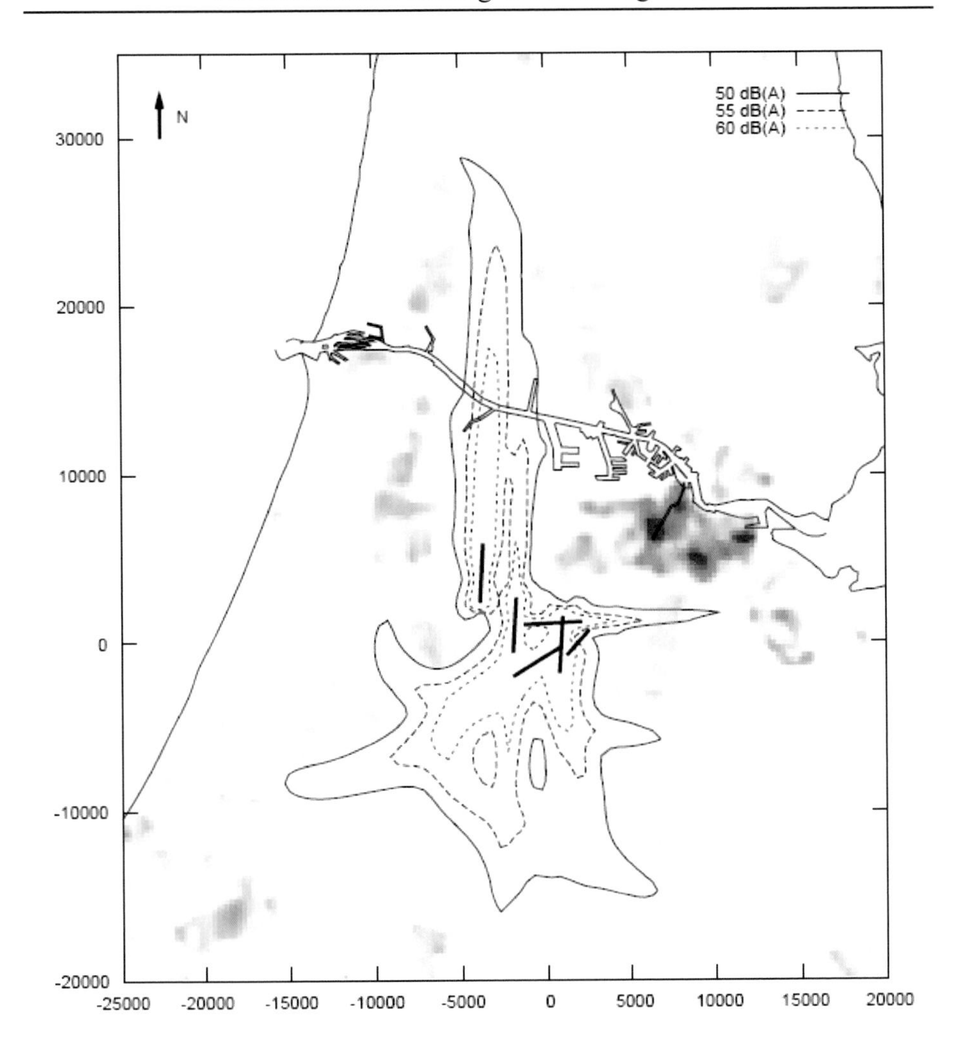

Figure 20. Reference and optimized noise results in L_{den} for Amsterdam Airport.

Figure 20 shows the graphical results with respect to noise for the metric L_{den}. The left part shows the reference scenario, the right part the optimized result corresponding to situation as shown in the right part of Figure 20. At first sight, not much has changed, although differences can be identified. It seems that more traffic is handled in North-South direction in the optimized result. However, when comparing the results using the Miedema noise annoyance dose response relationship, the potential benefits of using optimization become clear. The expected number of people annoyed by aircraft noise goes down from 98,100 for the reference scenario, to 69,000 for the optimized one. A similar reduction in risk is achieved. The sum of the individual risk experienced by all inhabitant of the shown area reduces from 0.048 to 0.035 casualties per annum. This means that both indicators have dropped by almost 30%.

Please note that SNAP calculates ideal annual average runway usage, i.e. it only calculates total runway use percentages for a full year, while assuming average weather and no traffic disruptions. It is therefore that it classifies as a tool in strategic range. At this point, it is not suitable for being employed at the operational level. When airport management has opted for a certain strategy and its resulting runway use, a different tool or possibly an extension to SNAP would be required to aid the airport or the responsible air traffic control authority in managing the strategic result at the operational level. Of course, such a tool could be part of the envisioned DSS for integrated environmental management.

Chapter 5

ENVIRONMENTAL PLANNING AT THE TACTICAL/OPERATIONAL LEVEL

NOISE ALLOCATION TOOL

Noise allocation is a process that does not aim for total reduction of noise load as experienced on the ground, but instead aims at a certain geographical allocation of the noise that is somehow beneficial for the communities surrounding the airport. We have already showed this principle in section V, through the creation of an optimized runway usage policy. However, aircraft routing can also influence the noise load distribution. Probably, route design is the most important factor in this, however route selection or allocation is also an important decision variable. In this section we will outline a planning tool that is based on this principle [Hebly, 2007].

For the implementation, an existing Mixed Integer Linear Programming (MILP) multi-runway sequencing and scheduling model for arriving traffic was used [Steijl, 1999]. In its original form, this model is already capable of selecting one of multiple fixed arrival routes to the airport for all inbound traffic. It solves for minimal delay, while respecting several separation requirements. The aim of the extension was to see whether a noise allocation policy could be added to the model and to investigate what the effects on the scheduling performance would be. With respect to the noise allocation policy, a noise sharing mode was selected for this problem.

MODEL DESCRIPTION

For the model, again Amsterdam Airport Schiphol is used as the example airport. One of main operating modes for this airport is to have the two parallel runways 18R and 18C in use for arrivals and runways 24 and/or 18L for departures. Currently most of the arriving traffic enters the Schiphol terminal area through one of the three available initial approach fixes (IAF). From there, traffic is vectored to one of the available arrival runways. Normally, traffic arriving from the west is assigned the 18R runway and traffic from the east lands at the 18C runway.

For this study, fixed arrival routes are defined from the three IAFs, as is shown in Figure 21. From each of them three different routes lead to the designated runway, where the middle route is considered the default one and the outer two as alternatives. From two of the IAFs here is also an extra route to the non-default runway. Concerning the profile, conventional step-down approaches are assumed using a 2,000 ft intercept altitude for the centre runway and a 3,000 ft intercept altitude for the right runway. For ease of interpretation with respect to the noise results, it is assumed that all aircraft are already at this altitude when crossing the IAFs. In reality, aircraft would be descending towards their intercept altitude at this point.

A wide mix of aircraft types visits the airport. Each aircraft has its own characteristics with respect to flight performance and noise production. In order to keep the model simple, three aircraft categories are used to represent the fleet, all three very different with respect to size, speed, noise level and separation requirements. The three models to represent the categories are the Boeing 747-400 (c1), the Boeing 737-400 (c2), and Embraer 120ER (c3). All of the required data for this aircraft is taken from or calculated by INM, see subsection II.c.

The optimization model is fed with an inbound traffic schedule, which specifies aircraft type, IAF and expected time over IAF. As soon as an aircraft crosses the IAF, it actually enters the model and starts the approach. Aircraft cannot be delayed in this process; however the model is allowed to delay the start of the approach. This type of delay is referred to as initial delay and should in reality be accomplished by delay during the transition or holding in a holding pattern.

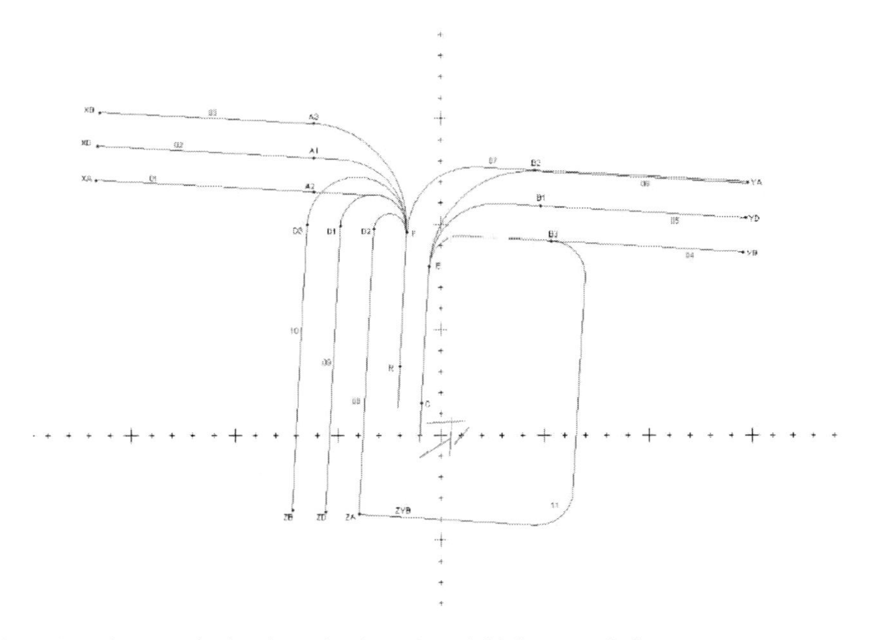

Figure 21. Eleven arrival trajectories from three initial approach fixes.

Because of the differences in aircraft performance and length of the arrival trajectories, the required time from IAF to the runway threshold differs. With three aircraft types and 11 possible routes, this results in 33 possible combinations of aircraft and routes, all with their distinct and predetermined travelling time. As already mentioned, no delay can be applied as soon as aircraft have started the approach. However, if the aircraft is scheduled to fly a non-default routes that takes more time than the default one, the difference is considered to be delay as well. This type of delay is labelled trajectory delay.

OBJECTIVES AND CONSTRAINTS

The chosen objective function for this problem minimizes the total impact of undesirable events. Three different types of events are penalized through this function: delay (initial as well as trajectory delay), the use of the non-default route and the use of the non-default runway. Mathematically, all events are modelled using integer variables, each counting the number of times a certain event takes place during the optimization period. Different penalty factors are used for each event, ranking them for undesirability. With respect to delay, the factors are chosen such that delaying a large aircraft is more expensive than delaying a small aircraft.

In order to get a valid solution from the model, 4 groups of constraints have been used. Each group is responsible for a certain type of behavior. They ensure that:

- all aircraft enter the system eventually, thus starting their approach,
- all aircraft keep moving along their trajectory with their predetermined speed once they have entered the system,
- all aircraft maintain a time-based separation measured at each waypoint with respect to other aircraft approaching over the same trajectory segment, where the applied separation is dependent on the weight class combination of the conflicting pair [Brinton, 1992],
- aircraft do not overtake each other in between to waypoints, for as far as the third group of constraints fails to prevent this.

Finally, there is a group of constraints related to the noise impact, to realize the chosen policy of spreading the noise as much as possible. The expected effect of this policy is that the area exposed to aircraft noise will increase, but that peak exposure levels will decrease. Such a policy might be chosen if noise exposure in a certain area in unacceptable high, or when noise exposure should be shared among different communities on a equitable basis.

In this model, the policy is implemented using constraints that control the difference between certain points. Nine of such points have been defined, indicated in Figure 21 with A_n, B_n and D_n. The constraints only limit the difference between the three points of a single group. For example, the difference between point A1 and A3 is controlled, but the difference between A1 and B1 is not. This prevents traffic from being redirected to another IAF because of noise spreading considerations, which is considered undesirable behaviour.

The maximum noise load difference allowed is equal to the difference cause by one approach of the noisiest aircraft in the actual traffic sample for each of the entry points. This rule makes sure that the differences between the three points of a single group will always be small, even if there are no approaches of the noisiest type, which is the B747-400 in this model.

OPTIMIZATION RESULTS

The effects of the noise constraints are shown using two optimization runs. The first result is calculated using the model, but without the noise sharing constraints enabled. The solution is than compared to the results of a run with the same input but where the noise allocation constraints are allowed to be active. Both optimizations solve for a one hour period, during which 31 aircraft arrive at the airport. Fourteen of them are of the c1, 11 of the c2 and 6 of the c3 category. Twenty of the 31 aircraft arrive in the first half hour of the optimization run, which leaves the second half rather slow with only 11 flights remaining.

To start with, the noise results of both runs can be compared. They are depicted in Figure 22. The left part of Figure 22 represents the situation with the noise constraints disabled. Comparing both results, it can be observed that for the noise sharing mode the peak levels are – as expected - reduced at least one colour band for all three regions. However, the reduced peak levels appear in a wider area. When observing the right part of Figure 22 only, two other observations can be made. The first one is that the solver does indeed allow for differences between the three regions. Second, the noise load from the traffic from the south and the east IAF is still spread somewhat unequal over their respective region. This is not only allowed to give the solver some flexibility with respect to the delay optimization, it may also be required in obtaining a valid solution, as the traffic schedule may simply not offer the traffic that can be distributed equitably with respect to noise.

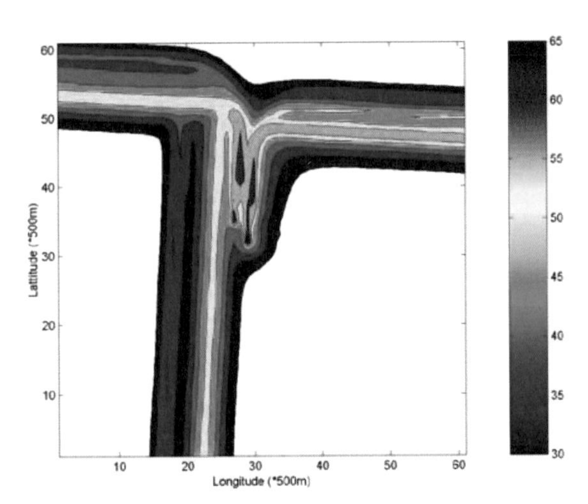

Figure 22 (Continued)

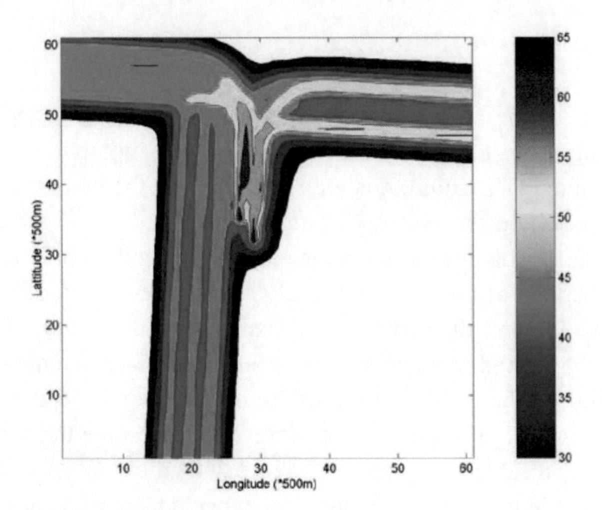

Figure 22. The noise results for both optimization runs, on the left without the constraints and on the right with the noise allocation constraints. Both results show LA_{eq} for a one-hour period.

To start with, the noise results of both runs can be compared. They are depicted in Figure 22. The left part of Figure 22 represents the situation with the noise constraints disabled. Comparing both results, it can be observed that for the noise sharing mode the peak levels are – as expected - reduced at least one colour band for all three regions. However, the reduced peak levels appear in a wider area. When observing the right part of Figure 22 only, two other observations can be made. The first one is that the solver does indeed allow for differences between the three regions. Second, the noise load from the traffic from the south and the east IAF is still spread somewhat unequal over their respective region. This is not only allowed to give the solver some flexibility with respect to the delay optimization, it may also be required in obtaining a valid solution, as the traffic schedule may simply not offer the traffic that can be distributed equitably with respect to noise.

The solutions can also be compared with respect to flying time and delay. These results are given in Table 3.

With respect to the total flying time (including initial delay), the difference is only 2.5 minutes, which is less than 1%. The noise constraints increase the trajectory flying time by 5 minutes, but at the same time decrease initial delay for this sample problem. Especially the c1 category faces increased flying time, because of more non-default trajectory allocations. This can be explained from the fact that spreading the noisiest aircraft type is the most efficient way of spreading the noise load. However, this apparently also gives the opportunity to allocate the smaller aircraft more efficiently, because combined their trajectory flying time decreases from 166.5 to 164 minutes. One should however keep in mind that although the difference in time is less than 1%, the difference in cost is not necessarily less than 1% as well. In fact, since it is the largest aircraft type that is delayed, this is probably not the case.

Table 3. Duration of arrival trajectories and imposed delay

All times in minutes	Noise sharing disabled	Noise sharing enabled
Trajectory flying time		
C1	116	121
C2	95	93.5
C3	65	66.5
Total trajectory	276	281
Initial delay		
C1	0.5	0.5
C2	3	2
C3	3.5	2
Total initial	7	4.5
Total all	283	285.5

Concluding, this study shows that it is perfectly possible to add noise related objectives to an initially delay driven sequencing and scheduling problem for arriving flights. The chosen approach results in a more equal sharing of the noise, decreasing the highest cumulative noise levels at the cost of increased areas with moderate exposure, without a very large impact on the efficiency of the scheduling process.

CURRENT AND FUTURE WORK

The previously developed noise allocation planning tool is not really a real-time decision support system, but merely an academic planning model. It offers a good possibility to study the effects of including noise allocation considerations into the scheduling problem. At the same time, the tool is focussed on distributing or sharing noise at a particular point in order to reduce peak exposure. Although it was shown that this approach works quite well, other noise allocation policies might be pursued. For example, policies that take into account population density may help in reducing total noise annoyance.

Current research focuses on a similar topic, but with two clear distinctions. First of all, the focus will be to integrate the routing preferences for the different aircraft into an arrival manager, an already existing real-time decision support system used for sequencing and scheduling arriving traffic. Second, a more flexible approach will be used with respect to the noise allocation, allowing for different noise allocation policies based on stakeholder demand or governmental regulation. This should also include the possibility to use multiple objectives with respect to the noise exposure based on different time frames. This should allow for a situation where for example both the current day as well as the cumulative annual noise exposure is taken into consideration.

With respect to future research, more focus will be on the tactical/-operational level, especially on how to integrate and thereby enable the developments at the trajectory and the strategic level in the air traffic control process. This should lead to the envisioned decisions support system, which may become available in multiple levels of functionality. Eventually, in its final form and in an ATC environment of trajectory based operations, it should support controllers in maintaining spacing of the individual flights that fly custom, environmentally optimized trajectories.

CONCLUSION

In this Chapter, our research efforts directed towards the development of an integrated airport environmental management system have been summarized. The set of fully integrated noise management tools that we envision, includes interrelated tools at the strategic level (annual/seasonal noise allocation planning), the tactical-/operational level (sequencing and scheduling of flights and separation assurance), and the trajectory level (noise-optimized routes and flight paths). The core element in the proposed concept, which is at the tactical-/operational level, is an air traffic controller decision support tool that helps controllers in their routing, sequencing and scheduling tasks, with the aim to maximize throughput efficiency in concert with noise exposure considerations.

To ensure that the cumulative, long-term noise impact is reduced, the proposed operational tool, which aims at helping controllers to distribute the traffic flows over noise sensitive areas, is coupled to the strategic noise allocation planning tool, which seeks to ascertain the maximum annual number of aircraft operations that can be accommodated within the regulatory limits for average annual noise exposure. To this end, the short-term noise level targets set in the operational tool for the different day-to-day airport operating arrangements are directly derived from the long-term noise objectives established by the strategic noise allocation planning tool.

In the concept that we envision the need for low-altitude vectoring and holding is completely eliminated by resorting to a fundamental change in the way air traffic is controlled. In our concept, air traffic controllers establish efficient control and spacing of individual flights through the operational management of environmentally optimized 4D trajectories. This enables the use of low-noise, reduced-emissions arrival and departure flight paths, not only in periods of low demand, but also during peak hours.

Although the main enabling capabilities (tools) that underlie the proposed integrated environmental management concept are still in various stages of development, and the actual integration of the tools at the various levels of aggregation is yet to take place, there are at present no indications that appear to refute the promise for major holistic environmental impact reductions that the concept holds out. Continued development of this promising concept will therefore be vigorously pursued.

ACKNOWLEDGMENTS

The authors gratefully acknowledge the support of several MSc students of the TU Delft, notably, Hans Steijl, Roel Hogenhuis, Robin van Loo, Victor Hanenburg and Edwin de Bruijn.

REFERENCES

Airports Council International. ACI Airport Economics Survey –2001; Geneva, Switzerland.

Anton Cruz, M., et al.. *Environment Requirements and Operational Constraints;* Report SOURDINE/AENA-D1-DOC-001, 2000.

Brinton, C.R. An Implicit Enumeration Algorithm for Arrival Aircraft Scheduling. *Proceedings of the IEEE/AIAA Digital Avionics Systems Conference,* Seattle, WA, U.S.S. 1992.

Clarke, J.P. and Hansman, R.J. Systems Analysis of Noise Abatement Procedures Enabled by Advanced Flight Guidance Technology. *Journal of Aircraft, Vol.* 37, No. 2, 2000, pp.266-273.

Committee on Aviation Environmental Protection (2007). *Airport Air Quality Guidance Manual.* http://www.tc.gc.ca/civilaviation/International/ ICAO/committee/pdf/working/ Jan-22- 07/CAEP7_WP28.pdf.

Davison, H.J.; Reynolds, T.G. and Hansman, R.J. Human factors implication of continuous descent approach procedures for noise abatement. *Air Traffic Control Quarterly*, Vol 14, No 1, 2006, pp.25–45.

Deb, K. *Multi-Objective optimization using Evolutionary Algorithms.* Wiley, Chichester, 2001.

Department of Transport and Regional Services (2003). Guidance material for selecting and providing aircraft noise information. http://www.dotars.gov.au/aviation/environmental/ transparent_noise/guidance/index.aspx.

Erkelens, L.J.J Research into New Noise Abatement Procedures for the 21st Century, AIAA-2000-4474, Proceedings of the AIAA Guidance, Navigation, and Control Conference, Denver CO, U.S.A., 2000.

EUROCONTROL (2005). Third Party Risk. http://www.eurocontrol.int/environment/public/ standard_page/third_party.html

EUROCONTROL Experimental Centre (2006). Noise and Emission Modelling methodology.http://www.sourdine.org/documents/public/SII_WP5_D5-2_v10.pdf.

European Commission (2002). Position paper on dose response relationships between transportation noise and annoyance. http://ec.europa.eu/environment/noise/pdf/ positionpaper.pdf.

Eagan, M.E. (2006). Using Supplemental Metrics to Communicate Aircraft Noise Effects. http://www.trbav030.org/pdf2007/TRB07_Eagan-ppt.pdf

FAA Office of Environment and Energy. Integrated Noise Model (INM) Version 5.1 Technical Manual. Report FAA-AEE-97-04, 1997.

FAA Office of Environment and Energy (2007). Emissions and Dispersion Modeling System (EDMS) Version 5.0 User's Manual. http://www.faa.gov/about/ office_org/ headquarters_offices/aep/models/edms_model/media/EDMS% 205.0.2%20User% 20Manual.pdf

Federal Aviation Administration. Noise Control and Compatibility Planning for Airports. *Advisory Circular.* 150/5020-1, 1983

Federal Interagency Committee on Aviation Noise (1997). Effects of Aviation on Awakenings from Sleep. http://www.fican.org/pdf/Effects_AviationNoise_Sleep.pdf

Foot P.B., et al. Third Party Risk near Airports and Public Safety Zone Policy. *NATS, HMSO,* London, R and D Report 9636, 1997.

Hale, A. Risk contours and risk management criteria for safety at major airports, with particular reference to the case of Schiphol. *Safety Science,* Volume 40 (1-4), 2002, pp.299-323.

Hargraves, C.R. and Paris, S. W.; Direct trajectory optimization using nonlinear programming and collocation. *J. Guid. Control Dyn.,* Volume 10, No.4, 1997, pp.338–342.

Hebly, S.J. and Wijnen, R.A.A. Development of a Runway Allocation Optimisation Model for Airport Strategic Planning. *Proceedings of the Australian Aeronautical Conference, Melbourne,* Autralia, 2005.

Hebly, S.J.; Hanenburg, V; Wijnen, R.; and Visser, H.G. Development of a noise allocation tool. *Proceedings of the 45th Aerospace Sciences Meeting,* Reno, NV, U.S.A., 2007

Holland, M. and Watkiss, P. (2002). Estimates of the marginal external costs of air pollution in Europe. http://ec.europa.eu/environment/enveco/air/pdf/betaec02a.pdf

Hough, J.W. and Weir, D.S. (1996). Aircraft Noise Prediction Program (ANOPP) Fan Noise Prediction for Small Engines. http://ntrs.nasa.gov/archive/nasa/casi.ntrs.nasa.gov/ 19960042711_1996066507.pdf

International Civil Aviation Organization. ICAO Document 8168- OPS/611 — *Aircraft Operations,* Volume I, 1999.

Jelinek, F., Carlier, S. and Smith, J. (2004). Advanced Emission Model (AEM3) Version 1.5. http://www.eurocontrol.int/eec/gallery/content/public/documents/ EEC_SEE_reports/EEC_SEE_2004_004.pdf

Joint Planning and Development Office (2007). Concepts of Operations for the next generation air transportation system. http://www.jpdo.gov/library/NextGen_v2.0.pdf

Kershaw, A.D.; Rhodes, D.P. and Smith, N.A. The influence of ATC in approach noise abatement. *Proceedings of the 3rd USA/Europe Air Traffic Management R and D Seminar,* Napoli, Italy, 2000.

Maris, E,; Stallen, P.J.; Vermunt, R.; and Steensma, H. Noise within the social context: Annoyance reduction through fair procedures. *J. Acoust. Soc. Am.,* Volume 121, Issue 4, 2007, pp. 2000-2010.

Metron Aviation. Noise Integrated Routing System. http://www.metronaviation.com/ content/text/nirs.pdf, Accessed July 2007.

Miedema H.M.E. and Vos, H.; Exposure-response relationships for transportation noise. *J. Acoust. Soc. Am.,* Volume 104, Issue 6, 1998, pp.3432-3445.

Miedema H.M.E.; Vos, H. and De Jong R.G. Community reaction to aircraft noise: time-of-day penalty and tradeoff between levels of overflights. *J. Acoust. Soc. Am.,* Volume 107, Issue 6, 2000. pp.3245-3253.

Optimal. Outlook of the OPTIMAL project. http://www.optimal.isdefe.es/, Accessed July 2007.

Penner, J.E., Lister, D.H., Griggs, D.J., Dokken, D.J., and McFarland, M. (1999). Aviation and the Global Atmosphere. http://www.grida.no/climate/ipcc/aviation/index.htm.

Ruijgrok, G.J.J. *Elements of Aviation Acoustics*; Delft University Press, Delft, The Netherlands, 2004.

Steijl, H.. *Optimization of multi-runway sequencing and scheduling*. M.Sc. Thesis, Faculty of Aerospace Engineering, Delft University of Technology, Delft, The Netherlands, 1999.

Visser, H.G. and Wijnen, R.A.A. Optimization of Noise Abatement Departure Trajectories. *Journal of Aircraft*, Vol. 38, No. 4, 2001, pp. 620-627.

Visser, H.G. and Wijnen, R.A.A. Optimisation of Noise Abatement Arrival Trajectories. *The Aeronautical Journal*, Vol.107, No. 1076, 2003, pp. 607-615.

Visser, H.G. Generic and Site-Specific Criteria in the Optimization of Noise Abatement Trajectories. *Transportation Research Part D: Transport and Environment*, Volume 10, Issue 5, 2005, pp. 405-419.

Visser, H.G. Noise Optimal Resolution of Air Traffic Conflicts, *Proceedings of the 2006 INTERNOISE Congress, Honolulu, HI, U.S.A.,* 2006.

Wijnen, R.A.A. and Visser, H.G. Optimal Departure Trajectories with Respect to Sleep Disturbance. *Aerospace Science and Technology*, Vol.7, No. 1, 2003, pp.81-91.

Wijnen, R.A.A. Adopting the Agile Unified Process for Developing a DSS for Airport Strategic Planning. *Proceedings of the International Conference on Research in Air Transportation,* Belgrade, Serbia and Montenegro, 2006.

INDEX

A

abatement, ix, x, xiii, 16, 17, 21, 22, 23, 26, 28, 38, 39, 79, 81
access, ix, 17, 58
accidents, 11
accuracy, 18, 25, 26, 38, 39
adjustment, 1, 2
aggregates, 3
aggregation, xi, xiii, 15, 17, 19, 22, 76
agriculture, 58
air pollution, xii, 8, 51, 80
air quality, ix, xi, 8, 9
airlines, 19
airports, vii, ix, x, xi, 3, 7, 8, 12, 13, 15, 19, 57, 58, 60, 80
algorithm, 22, 29
alternative, xii, 7, 19, 31
alternatives, 4, 68
anxiety, xii
assessment, xiii, 3, 7, 9, 12, 22, 23, 55
assignment, 8
assumptions, 8
attention, 18
Australia, 3
authority, 64, 66

B

barriers, ix
behavior, 28, 31, 38, 41, 43, 53, 55, 70
benefits, x, xii, 17, 19, 22, 31, 38, 50, 65
blocks, xiii
building blocks, xiii
burn, 16, 21, 22, 23, 24, 29, 51

C

carbon, 9, 22
carbon monoxide, 9, 22
categorization, 15
cell, 24
certification, 1, 9, 11
climate change, 8
CO2, 10
community, x, xi, xii, 2, 3, 5, 6, 8, 17, 22, 23, 30, 42
compatibility, ix, 16, 58
compensation, 16
compilation, 24
complement, 17
compliance, 9
components, 7, 28, 40
composition, 31
computation, 7, 24
computing, 19
concentration, 11
confidence, 4
confidence interval, 4
configuration, 8, 26, 27, 38, 39, 60, 61
conflict, 16, 39, 54, 61
conflict resolution, 55
Congress, 82
constraints, x, 22, 25, 27, 28, 41, 53, 60, 61, 70, 71, 72
construction, 13, 57
consumption, 30, 41, 43, 50, 51, 54
control, x, xii, 16, 17, 18, 25, 26, 38, 58, 66, 70, 74, 75
costs, 51, 52
coverage, 38
crop production, 51

D

data collection, 3
data set, 3, 25
database, 8
decision making, xii, 22
decisions, 17, 54, 58, 74
demand, vii, ix, x, xi, xiii, 58, 61, 74, 75
density, 23, 24, 30, 38, 42, 58, 74
developed countries, 9
developed nations, 11
deviation, 31
differential equations, 25
discretization, 25
dispersion, 9, 38
distribution, xi, 12, 19, 21, 23, 24, 30, 42, 59, 60, 61, 67
dosage, 5
dose-response relationship, 3, 5, 6, 22, 24
duration, 2, 28, 61
dynamical systems, 26

E

ears, 1
economics, 58
efficiency criteria, 19
emission, vii, 9, 10, 15, 22, 23, 24, 52
emission source, 9
empowerment, xii
energy, 1, 2, 16, 43, 50
environment, xiii, 9, 18, 74, 80
environmental effects, 15, 17, 18, 59, 60
environmental impact, iv, vii, ix, xi, xiii, 15, 16, 18, 19, 22, 58, 76
environmental issues, x, 3, 17, 58
equilibrium, 24
equipment, 9
estimating, 10
Europe, x, 2, 3, 10, 80, 81
European Commission, 80
evening, 2
execution, 16, 17
expertise, 59
exposure, vii, xi, 2, 3, 5, 6, 7, 8, 17, 22, 23, 24, 30, 42, 60, 70, 73, 74, 75, 81
external costs, 51, 52, 80

F

FAA, 7, 9, 57, 80
failure, 28
fair procedures, 81
family, 2
fidelity, 7
financing, 59
first generation, 53
flexibility, 19, 38, 71, 72
flight, vii, x, xi, xii, xiii, 2, 7, 8, 9, 10, 11, 18, 22, 23, 24, 25, 26, 28, 30, 31, 38, 39, 41, 42, 43, 50, 51, 53, 55, 61, 68, 75
fluctuations, 1
focusing, xii, 15
fuel, 10, 16, 21, 22, 23, 24, 25, 28, 29, 30, 31, 33, 37, 41, 42, 43, 45, 49, 50, 51, 52, 54
fuel flow rate, 10

G

generation, 53, 81
Geneva, 79
global climate change, 8
goals, xii, 57
government, iv, 57, 58
groups, 70
growth, vii, ix, x, xi, 58
guidance, x, 12, 16, 38, 79
guidelines, 57

H

health, 2, 51
height, 9
hospitals, 57
housing, 58
hydrocarbons, 9, 22

I

impact assessment, xiii, 7, 23
implementation, 17, 19, 26, 67
indicators, 2, 7, 10, 66
indices, 10, 11
infrastructure, x
initial state, 27, 40
initiation, 41
instruments, 58
insulation, ix, 16, 19, 57, 58
integration, 25, 76
intensity, 1, 21, 43
interaction, 19
interactions, xiii
interface, 63
interference, 5
interpretation, 68
interrelationships, xiii
interval, 4, 25, 26
isolation, xi, 17
Italy, 81
iteration, 63

K

knots, 28

L

land, ix, x, 3, 12, 16, 57, 58, 59
land acquisition, 59
land use, 16, 57, 58
land-use, 3, 12
legislation, 9, 12
life cycle, 15
limitation, 3, 31, 50
literature, 1
location, 2, 8, 12, 29, 53, 57
lying, 72

M

management, vii, x, xi, xii, xiii, 17, 18, 19, 39, 58, 59, 66, 75, 76, 80
marginal external cost, 51, 52, 80
market, 58
measurement, 1, 3, 9
measures, ix, x, xi, 15
media, 80
MIP, 61
missions, 22
modeling, 7, 9, 19
models, 7, 8, 9, 59, 68, 80
modules, 59
Montenegro, 82
motion, 25, 27
movement, xi, 2, 7, 8, 12, 40, 42, 60

N

Netherlands, x, xi, 12, 13, 22, 52, 53, 58, 81
next generation, 81
nitrogen, 9, 22
nitrogen oxides, 9, 22
nodes, 25
noise, iv, vii, ix, x, xi, xii, xiii, 1, 2, 3, 4, 5, 6, 7, 8, 15, 16, 17, 19, 21, 22, 23, 24, 26, 28, 29, 30, 31, 37, 38, 39, 41, 42, 43, 49, 50, 51, 52, 54, 55, 57, 58, 59, 60, 61, 63, 65, 67, 68, 70, 71, 72, 73, 74, 75, 79, 80, 81
noise-optimized, vii, xii, 29, 30, 31, 42, 43, 52, 75
North America, 3

O

observations, 71, 72
optimal performance, 29, 42, 50
optimization, 4, 15, 19, 22, 23, 25, 28, 29, 30, 31, 40, 41, 43, 50, 51, 52, 53, 59, 60, 61, 62, 63, 64, 65, 68, 69, 71, 72, 79, 80

optimization method, 25
organization, 5
oxides, 9, 23

P

parameter, 31, 32, 53
Paris, 80
penalties, x, 58
perception, 1, 3, 6
performance, 4, 7, 8, 15, 16, 18, 19, 21, 22, 23, 24, 25, 28, 29, 30, 31, 38, 39, 41, 42, 43, 50, 51, 53, 55, 58, 67, 68, 69
permit, 22
phase transitions, 26
pitch, 1
planning, vii, xii, xiii, 3, 12, 16, 18, 57, 58, 59, 67, 73, 75
plants, 57
pollutants, 9, 22, 51, 52
pollution, xii, 8, 9, 51, 80
polynomials, 4
poor, xi
population, 4, 5, 6, 11, 21, 23, 24, 30, 42, 60, 74
population density, 24, 30, 42, 74
potential energy, 50
power, 8, 10, 11, 19, 42
predictability, xi
prediction, 7, 18
prediction models, 7
pressure, 1, 2
prices, 52
probability, 6, 11, 12
production, 51, 68
program, 7, 19
programming, 25, 80
propagation, 7, 8, 21

Q

quality of life, ix

R

radar, 16, 38
radius, 53
range, 1, 15, 18, 24, 29, 40, 41, 51, 57, 59, 61, 66
reality, 68
recognition, xii
reduction, xi, 15, 16, 18, 19, 21, 30, 31, 38, 41, 42, 43, 52, 58, 65, 67, 81
regional, 7
regulation, 74
regulations, xi, 9, 15, 57
relationship, xi, 5, 6, 22, 24, 65
relationships, 3, 5, 80, 81
resolution, 54
risk, ix, 11, 12, 13, 14, 57, 59, 60, 61, 63, 65, 80
risk assessment, 12
risk management, 80
routing, 19, 52, 60, 67, 74, 75
rural areas, 52

S

safety, vii, xi, 11, 13, 16, 17, 22, 59, 60, 80
sample, 70, 72
scheduling, vii, xii, 67, 73, 74, 75, 81
school, 57
sea level, 24
sea-level, 10
segmentation, 25, 26
selecting, 60, 67, 79
sensitivity, 1
separation, vii, xi, xii, 16, 38, 54, 67, 68, 70, 75
sequencing, vii, xii, 16, 67, 73, 74, 75, 81
Serbia, 82
series, 38
shape, x, 30, 42
shaping, 38, 58
sharing, 39, 67, 71, 72, 73
SII, 80
sleep disturbance, 5, 6, 24

smoke, 9
SNAP, xiii, 19, 59, 60, 61, 64, 66
social context, 81
software, 8, 25
sounds, 1, 5
speed, ix, 27, 28, 30, 38, 39, 40, 41, 50, 52, 53, 55, 68, 70
stages, 76
stakeholders, 58
standards, xi, 9
strategic planning, 58, 59
stretching, 50
students, 77
sulphur, 23
Switzerland, 79
symptoms, xii
synthesis, xiii, 19, 22, 23
systems, 26, 38, 53

T

tactics, 16
targets, 38, 75
technological developments, 17
technology, xi, 15, 39, 52
temperature, 21
three-dimensional space, 30, 42
threshold, ix, xi, 2, 28, 40, 69
time constraints, 23, 39
time frame, 74
timing, 59
trade, 7, 17, 22, 29, 61
trade-off, 7, 17, 22, 29, 61
traffic, vii, ix, x, xi, xii, 9, 11, 16, 17, 18, 38, 60, 61, 64, 65, 66, 67, 68, 70, 71, 72, 74, 75
trajectory, vii, xii, xiii, 7, 15, 16, 17, 18, 19, 22, 23, 24, 25, 26, 27, 28, 29, 30, 31, 38, 39, 40, 42, 43, 50, 51, 52, 53, 54, 69, 70, 72, 73, 74, 75, 80
transition, 18, 31, 68
transitions, 26, 31, 43
transport, 3, 22
transportation, 80, 81

U

UK, 12, 13
uncertainty, 38
uniform, 18, 19
United States, 7
urbanization, ix
users, 62

V

values, 4, 10, 25, 28, 29, 31, 40, 41, 42, 51, 62, 63
variable, 67
variables, 22, 25, 26, 40, 69
vector, 28, 40
vision, 16

W

weakness, 6
welfare, 2
wind, 7, 18, 21, 25, 55, 61

yield, 61